Russian Airliners

Outside Russia

Dmitriy Komissarov and Yefim Gordon

MIDLAND
An imprint of
Ian Allan Publishing

Russian Airliners Outside Russia

ISBN (10) 1 85780 252 7
ISBN (13) 978 1 85780 252 8

Published by Midland Publishing
4 Watling Drive, Hinckley, LE10 3EY, England
Tel: 01455 254 490 Fax: 01455 254 495
E-mail: midlandbooks@compuserve.com

Midland Publishing is an imprint of
Ian Allan Publishing Ltd

Worldwide distribution (except North America):
Midland Counties Publications
4 Watling Drive, Hinckley, LE10 3EY, England
Telephone: 01455 254 450 Fax: 01455 233 737
E-mail: midlandbooks@compuserve.com
www.midlandcountiessuperstore.com

North American trade distribution:
Specialty Press Publishers & Wholesalers Inc.
39966 Grand Avenue, North Branch, MN 55056
Tel: 651 277 1400 Fax: 651 277 1203
Toll free telephone: 800 895 4585
www.specialtypress.com

Design concept and layout
© 2007 Midland Publishing and
Polygon Press Ltd (Moscow, Russia)

Printed in England
by Ian Allan Printing Ltd
Riverdene Business Park, Molesey Road,
Hersham, Surrey, KT12 4RG

Acknowledgements
The authors wish to express their sincere thanks to the people who assisted in making this book: Viktor Kravchenko, Sergey Krivchikov, Dmitriy Petrochenko and Yuriy Kirsanov, who supplied many of the high-quality photos used in this book; Vnukovo Airport Aviation Security and East Line Aviation Security which granted ramp access, allowing valuable photos to be taken; Lidia Anghelova, Natal'ya Titova and the late Ivan Faleyev at the CIS Interstate Aviation Committee; the editorial team of Soviet Transports (Peter Hillman, Stuart Jessup, Guus Ottenhof and Tony Morris), as well as Nikolay Ionkin, for supplying hot information on aircraft movements and construction numbers; Nigel Eastaway of the Russian Aviation Research Trust, whose extensive archive proved very useful; and, last but not least, Sergey and Andrey Samoylov for their dedication, perseverance and a lot of overtime work.

Title page: **The 'twin arrow' cheatline on AZAL Cargo's IL-76TD 4K-AZ16 (c/n 1023412411, f/n 8603) reveals previous ownership by Avialeasing of Uzbekistan, with which it served as UK 76410.** *Yefim Gordon archive*

Opposite page, top: **An immaculate Civil Aviation Administration of China (CAAC) Ilyushin IL-14P registered 648 (c/n 146000919) prepares to take on a load of passengers at Kunming.** *Philip Beer*

Opposite page, bottom: **MIAT Mongolian Airlines Tu-154M MPR-85644 (c/n 88A780).** *via Yefim Gordon*

Contents

Introduction

No sooner had the Soviet Union established itself as an aviation power in the late 1920s than it began seeking foreign customers for aircraft of Soviet design and manufacture. The modest export deliveries achieved in the years prior to the Second World War were almost entirely concerned with military aircraft; however, the first precedent was created in those far-away days when the Soviet government donated a Tupolev ANT-9 2M-17 twin-engined nine-seat airliner to the Turkish government.

Exports of Soviet commercial aircraft began in earnest after the war when the Soviet Union started expanding its influence and building the Eastern Bloc. With the Cold War in progress, the nations on the east side of the 'Iron Curtain' were not allowed to buy American or West European airliners (with a few exceptions) and, *volens nolens*, had to rely on the USSR to help rebuild their air transport systems. Thus Bulgaria, the People's Republic of China, Czechoslovakia, East Germany (the German Democratic Republic), Hungary, Romania and Yugoslavia became the first recipients of Soviet civil aircraft in the late 1940s and early 1950s when such airlines as TABSO Bulgarian Air Transport, CAAC, ČSA Czechoslovak Airlines, Deutsche Lufthansa (the *East German* one – the precursor of Interflug), Malev and TARS were born (or reborn). New export possibilities opened up when the colonial system broke down in the 1960s, with numerous African states gaining sovereignty and facing the tough choice of a sociopolitical development system; the Middle East, Asia and Latin America were in a turmoil and facing similar choices. The Soviet Union put this situation to good use, 'fishing in murky waters' – that is, trying to win the developing countries over into the socialist bloc and thus establish new outposts of Soviet polliltical and military power around the world. The means used to achieve this included deliveries of civil and military hardware on very favourable terms and helping the newly independent countries build their transport systems from scratch.

This was the time when the traditional markets for Soviet and Russian commercial aircraft were formed. Early exports were modest, such aircraft as the Ilyushin IL-12 and the first Soviet jet airliner, the Tupolev Tu-104, being sold abroad in extremely limited numbers. However, more customers were found as the Soviet aircraft industry brought out more advanced products and the operators became more familiar with them. The Antonov An-24/An-26/An-30/An-32 twin-turboprop regional airliner/transport family, the Ilyushin IL-76 transport and the Tupolev Tu-134 and Tu-154 medium-haul airliners became best-sellers among Soviet commercial aircraft – not to mention the 'old but gold' An-2 utility biplane which proved so capable and indestructible that production continued all the way until 2004!

In fairness, the Soviet aircraft design school and the ability of Soviet aircraft to do the job they were designed for earned universal recognition and even attracted orders from 'the wrong side of the Iron Curtain'. The diminutive Yakovlev Yak-40 trijet saw airline service in West Germany and Italy – albeit briefly.

Came 1991, and political unrest in the Soviet Union put an end to its existence; the Commonwealth of Independent States was formed instead by mutual consent. Thus, the one-time all-Union airline Aeroflot disintegrated into hundreds of airlines big and small. Many of the constituent Civil Aviation Directorates (CADs), several of which were situated in the Russian Federation and one in each of the other Soviet republics, suddenly found themselves abroad, forming the basis for new national flag carriers.

It has to be said that Aeroflot's organisation closely resembled an air arm's order of battle – which is hardly surprising, considering that the Soviet civil air fleet constituted an immediately available military reserve. The CADs were broadly equivalent to the air forces of the USAF or the air armies of the Soviet Air Force. Each CAD comprised several United Air Detachments (UADs) equivalent to an air group (USAF) or an air division (SovAF). In turn, each UAD consisted of several Flights similar to an air wing (USAF) or an air regiment (SovAF) and comprising several squadrons. Quite often the UADs became separate airlines in post-Soviet times.

Over the years, Soviet/Russian commercial aircraft have seen service in dozens of countries around the world. In Europe, new and second-hand aircraft have been delivered to, inherited by or leased by Armenia, Azerbaijan, Belarus, Bulgaria, former Czechoslovakia (the Czech Republic and Slovakia), Estonia, Georgia, (East and West) Germany, Hungary, Italy, Latvia, Lithuania, Moldova, Poland, Romania, Sweden, Switzerland, the Ukraine, the UK and the former Yugoslavia (Bosnia & Herzegovina, Macedonia and Slovenia).

In Asia, they have seen service in Afghanistan, Albania, Bangladesh, Cambodia, the People's Republic of China, India, Indonesia, Iran, Kazakstan, Kyrghyzstan, North Korea, South Korea, Laos, Mongolia, Nepal, the Philippines, Sri Lanka, Tajikistan, Turkey, Turkmenistan, Uzbekistan and Vietnam.

In the Middle East, Soviet commercial aircraft were operated by Jordan, Lebanon, Syria. African nations operating (now or previously) Soviet and Russian commercial aircraft are Angola, Benin, Burkina Faso, Burundi, the Central African Republic, Congo-Brazzaville, Djibouti, Egypt, Equatorial Guinea, Ethiopia, Gambia, Ghana, Guinea, Liberia, Libya, Madagascar, Mali, Mauritania, Mozambique, Nigeria, Sao Tome & Principe, Rwanda, Sierra Leone, South Africa, Sudan, Swaziland, Tunisia, Yemen (North and South), Zaire (now Democratic Republic of the Congo) and Zimbabwe.

In North America, Russian aircraft have seen commercial service in Canada and the USA (!); in Central and South America they have been operated by Bolivia, Brazil, Colombia, Cuba, Ecuador, Guyana, Honduras, Nicaragua, Panama, Peru, Uruguay and Venezuela. In Oceania, it is Papua New Guinea.

This album is an attempt to give an overview of Soviet/Russian-designed aircraft in service outside modern Russia – that is, the CIS republics are included, too. It Of course, not all countries operating Russian types commercially can be included for reasons of space. It should be noted that the album deals only with civil aircraft operations; that is, Soviet (Russian) transports delivered to the air forces of 'friendly nations' are not included – unless they are occasionally operated for the airline whose colours they nominally wear. As a rule, only the aircraft operated in a given country during the last 15 years or so are included for reasons of space. Helicopters are not included for the same reason – although mention of early deliveries and/or helicopters will be made in individual airline histories. Fleet rundowns given here are as up-to-date as possible but based on available information (details of aircraft ownership are often unavailable and airline fleets change fast).

Afghanistan

Afghanistan had operated Soviet transports since the 1960s, but deliveries began in earnest when Najibullah's Communist government came into power. **Bakhtar Afghan Airlines** took delivery of such aircraft as the Antonov An-24RV (two delivered), An-26 (five), Tupolev Tu-154M (two) and Yakovlev Yak-40 (three plus one bought from Czechoslovakia). Many of these aircraft were lost due to enemy action during the prolonged civil war. In 1988 the airline reverted to its original name, **Ariana Afghan Airlines [FG/AFG]**.

Additional Antonov aircraft were procured after the fall of Najibullah, including a pair of An-12 freighters and seven An-24s. Regrettably, again most of these were destroyed in US air raids during Operation *Enduring Freedom* in 2001. At present only two An-24RVs remain on the list, as Ariana has switched to Western types.

Ariana Afghan Airlines

Seen here on short finals to Moscow-Sheremet'yevo, Tupolev Tu-154M YA-TAR (c/n 87A748) displays its new Ariana Afghan Airlines titles. This aircraft was sold to Caspian Airlines of Iran as EP-CPG in March 1998; sister ship YA-TAP was less fortunate, being destroyed by shelling at Kabul in August 1992. *Yuriy Kirsanov*

Ariana also operated a pair of Antonov An-12 freighters, including An-12BP YA-DAA (c/n unknown) built in commercial configuration with no tail gunner's station; it is seen here visiting Sharjah. Both aircraft were destroyed by US air strikes at Kabul in November 2001 during Operation Enduring Freedom. *Yefim Gordon archive*

Albania

Ada Air Albania

Established in 1991, Ada Air Albania [ZY/ADE] operates scheduled passenger services to destinations in the Balkan Area. Although the current fleet consists of a single Embraer EMB-110P2 Bandeirante, the carrier periodically leases Yak-40 and Yak-42 trijets from Bulgaria and the Ukraine as required. The latter type was used on the Tirana-Zurich service.

Ada Air Albania leased this Yakovlev Yak-40, LZ-DOM (c/n 9620447), from Hemus Air thrice – in 1995-97, in 1999 and in 2005-06. The minimalistic white livery was obviously chosen for cost reasons; the eagle is the national symbol of Albania. *via Martin Štepanek*

Yak-42 UR-42377 (c/n 4520421014479, f/n 0711) leased from Donbass Airlines in 2000 had a slightly more fanciful livery with black trim and an Albanian flag; note the difference in the presentation of the titles. It is seen here during one of its visits to Zurich-Kloten.
Yefim Gordon archive

Angola

Airjet

An obscure Angolan air carrier called Airjet operated this Yak-40 registered D2-FEE and named *Junior* in 2003-04; its previous identity and ownership remain unknown. The aircraft is seen here undergoing maintenance at Sharjah in 2004. *via Bob Archer*

ALADA

ALADA – Sociedad de Transported [–/RAD] was established in 1995, flying passenger and cargo charter services from Luanda throughout Angola and to other African destinations. The current fleet consists of three An-32s and three IL-18s, all of which are operated in combi configuration. Additionally, a total of eight An-12 freighters and a single An-24 feederliner were operated previously at different times.

An-32 D2-FAX (ex-RA-48115, c/n 1510) in the standard livery of ALADA; it was later named *Kimoka*. Note the last letter of the registration repeated in extremely large format on the nose. *Yefim Gordon archive*

Another ALADA aircraft, Ilyushin IL-18E D2-FDY (ex-UR-75850, c/n 185008503), was bought from the Ukrainian carrier Kryla in 2001; the blue tail is a leftover from the ex-owner's livery.
Yefim Gordon archive

National Commuter Airlines

The trusty An-12 saw service with several Angolan air carriers, as well as the Angolan People's Air Force. Anonymous-looking commercial-configured An-12BP D2-FDB (c/n 02348207) seen at Sharjah in March 2000 was one of several owned by National Commuter Airlines. Later this aircraft became 3C-QQY.
Nikolay Ionkin

Armenia

Armenian Airlines

Armenian Airlines [R3/RME], the original Armenian flag carrier, were formed in 1991 as the successor of Aeroflot's Armenian CAD (comprising the 1st Yerevan UAD/ 113th Flight at Erebuni airport, the 2nd Yeveran UAD/297th Flight at Zvartnots and an independent air squadron at Gyumri). The carrier flew scheduled domestic and international services with nine Tu-134As and ten Tu-154B/B-1/B-2s; the nine Yak-40 were reportedly transferred to sister company Ararat Airlines but wore AAL colours. Cargo charters were flown with single examples of the An-12BK and the An-32B. In 2004 Armenian Airlines shut down. Armavia, the new flag carrier, adopted a more pro-Western image, and most of the original Soviet-built fleet found itself unwanted.

Tu-134A EK-65650 (c/n 0351006) in early AAL colours takes off from runway 24 at Moscow-Vnukovo in the mid-1990s. It was one of the airline's oldest aircraft.
Dmitriy Petrochenko

Convertible Yak-40K EK-88262 (c/n 9711752) in early Armenian Airlines colours is seen at Mineral'nyye Vody in February 1997. The AAL nose titles, like the tail logo, were in the Armenian flag colours; note that the Armenian Air Lines titles are repeated in Armenian script on the rear fuselage. *Dmitriy Petrochenko*

Armenian Airlines also flew cargo operations with this An-32B freighter, EK-48026 (c/n 3209), in full old AAL livery. *via Viktor Kravchenko*

Displaying Armenian Airlines' late livery, Tu-154B-2 EK-85566 (c/n 82A566) has just landed on runway 24 at Moscow-Vnukovo. *Dmitriy Petrochenko*

EK-86118, one of Armenian Airlines' two IL-86s (c/n 51483209086), sits on the cargo apron at Moscow-Domodedovo. It was sold to Air Van, another Armenian carrier, in 2005 but resold to Armavia that year (see next page). *Yefim Gordon archive*

Armavia

Armavia [U8/RNV] started out on a small scale with this Tu-134A, EK-65575 (c/n [93]62350, f/n 5209), in 2001; it is seen here on short finals to Moscow-Domodedovo. The aircraft retains the basic colours of Chernomorskie Airlines with whom it served as RA-65575. It is no longer in use. *Yefim Gordon archive*

Still in basic Armenian Airlines colours, IL-86 EK-86118 taxies towards the apron after landing on Domodedovo's runway 14L; note the new-style titles (in English and Armenian) and tail logo. *Sergey Krivchikov*

Ararat Avia

An-12BP EK-11351 (ex-Ukrainian Air Force UR-11351, c/n 4341910) is shown here in the smart livery of Ararat Avia in mid-1998; the tail logo is a stylised image of Mount Ararat. Note the logo of the Dvin Concern (the machine's previous owner) with which the freighter was jointly operated. *Yefim Gordon archive*

Armavia
Արմավիա
EK-86118

ARARAT AVIA
AN 12
ARMENIA
CARGO
EK-11351
DVIN
CONCERN

Dvin Air

Dvin Air [–/DVN] (also referred to as Dvin-Avia), the flying division of the Dvin Concern based at Yerevan-Erebuni, was chiefly a cargo carrier with a fleet of An-12s (seven of these are known to have been operated). However, the airline also flew passenger charters with a couple of Yak-40s, one of which, EK-87359 (ex-Qeshm Air EP-TQA, ex-Moldavian Airlines ER-87359, c/n 9340831) is seen here at Sharjah in 1999. The basic colour scheme is that of the Russian airline Trans-Charter. EK-87359 was leased to the Kenyan carrier African Air Express in 1999 and eventually sold to Armenian Airlines in 2002. *Peter Davison*

Likewise still wearing basic Trans-Charter colours, Yak-40 'Salon 2nd Class' EK-87662 (c/n 9240625) had different titles (in English and Armenian) and a larger logo. A curious feature of many Armenian aircraft is that the legend 'Armenia' was carried below the flag so as to preclude any doubts! This aircraft was likewise leased to African Air Express in 1999, later becoming EP-TQG with Qeshm Air in 2003 and finally 87662 with the Armenian government flight in 2006. *Yefim Gordon archive*

Dvin-Air also operated a single IL-76TD, EK-76445 (ex-Lasare Air 4L-76445, ex-Novosibirsk Airlines RA-76445, c/n 1023410330, f/n 8403) in 2000; the cheatline is that of Novosibirsk Airlines. The aircraft reverted to RA-76445 after sale to Gazpromavia in 2001. *Yefim Gordon archive*

Azerbaijan

After the collapse of the Soviet Union, Aeroflot's Azerbaijan CAD was transformed into the **Azerbaijan Airlines** or **AHY – Azerbaijan Hava Yollari** concern established on 7th April 1992. This comprised several enterprises. The Baku UAD comprising the 107th Flight (operating Tu-154B/B-1/B-2/Ms), 339th Flight (operating Tu-134B-3s and Yak-40s) and 360th Flight (operating An-26s). became **AZAL Avia [J2/AHY]**, operating scheduled and charter domestic and international services from Baku-Bina airport (now called Baku-Geydar Aliyev International Airport after post-Soviet Azerbaijan's first president who died in 2004). AZAL Avia had a total of 19 Tu-154s, eleven Tu-134Bs, 13 Yak-40s, three An-26Bs and two An-32Bs. Some Tupolev jets were re-export aircraft from Latvia.

The Zabrat UAD operating Mi-8 helicopters (109th Flight) and Mi-2 helicopters became AZAL Aero [–/AZK], performing utility work (including offshore oil support) and crop-spraying. The 50-odd An-2 crop-sprayer and utility biplanes were taken over by the AZAL Agro division based at Yevlakh. Later, in 1998, the AZAL Cargo division was formed and is described separately.

Originally AZAL's aircraft wore basic Aeroflot colours without titles or logo and no country prefix, their ownership being indicated only by the Azeri flag on the nose or tail. In 1993 the airline adopted its first own livery which existed with subtle variations until the introduction of the current blue/white livery at the turn of the century.

In the mid-1990s AZAL Avia embarked on a fleet modernisation/Westernisation programme, purchasing several well-used Boeing 727-200s. Work in this direction continued steadily, and in 2006 the only Russian aircraft remaining in the fleet were three Tu-134B-3s and four Tu-154Ms.

Azerbaijan Airlines

Tu-134B-3 4K-65714 (c/n [03]63527, f/n 6008) climbing away from Moscow-Domodedovo displays the oldest livery – standard except for the AZERBAIJAN titles (in Azeri) instead of AZERBAIJAN AIRLINES. Despite these titles, it is not a government VIP jet. *via Viktor Kravchenko*

In addition to the aircraft inherited from the Baku UAD, AZAL operates a few re-export aircraft, including Tu-134B-3 4K-65712 (c/n [03]63515, f/n 6006) seen landing on runway 14L at Moscow-Domodedovo. The aircraft retains the basic colours of Latavio Latvian Airlines with whom it flew as YL-LBL. *Yuriy Kirsanov*

In the early days of AZAL its aircraft wore basic Aeroflot colours and an Azeri flag on the tail, like this Yak-40 (4K-88174, c/n 9621247); note the small AZAL titles on the engine nacelle. Note also the AZAL An-26B and the Azerbaijan government (Air Force) IL-76M in the background. *via Martin Štepanek*

By the late 1990s the original livery had been livened up a bit by using more succulent colours, exemplified by Tu-154B-2 4K-85548 (c/n 82A548). Some aircraft carry the logo on an all-white tail. *via Viktor Kravchenko*

Tu-154M 4K-85734 (c/n 92A916) taxying at Domodedovo illustrates the old colours. It has now been reregistered 4K-734 and the original registration has resurfaced in the meantime as RA-85734, an ex-Chinese Tu-154M operated by Dalavia Far East Airlines. *Dmitriy Petrochenko*

AZAL Cargo

Azerbaijan Airlines' cargo division is a separate entity with its own flight code [–/AHC]. It operates An-26s, An-32s and IL-76TDs. In 2000 it briefly operated An-12BK UK 11418 No 1 (c/n 7344705) leased from Uzbekistan Airways; the aircraft wore the basic colours of Simurg, another Uzbek air carrier, with the AZAL-Cargo logo and AHC code on the tail. *Yuriy Kirsanov*

Imair

Baku-based Imair [IK/ITX], a division of the Improtex trade corporation, was established in October 1994. Scheduled and charter passenger and cargo flights were started in 1995. Much of Imair's fleet was leased from other carriers as required, such as the three IL-767MD freighters leased from Veteran Airlines (the Ukraine) and the An-12BPs leased from Ural Airlines in 1995-97. Here, Tu-134A EW-65049 (c/n [73]49755, f/n 3808) leased from Gomelavia taxies out for take-off from runway 32R at Moscow-Domodedovo in late 1998. The machine is in basic Gomelavia colours with Imair titles/logo and the Azeri flag. *Yuriy Kirsanov*

Over the years Imair operated six Tu-154s, mostly leased from Belavia. Wheels caught in mid-retraction after take-off from runway 14L at Domodedovo on a sunny winter day, one of Imair's two own Tu-154Ms, 4K-85732 (c/n 92A914), displays the carrier's attractive livery. *Dmitriy Petrochenko*

Silk Way Airlines

Silk Way Airlines [ZP/AZQ] is a cargo carrier founded in 2001, operating An-12s and IL-76s from Baku-Geydar Aliyev airport. The current fleet comprises two An-12BPs and six IL-76TDs. At least one of the latter is to be converted to IL-76TD-90 standard with PS-90A-76 turbofans; Silk Way Airlines is to be the second operator of this upgraded version after the Russian carrier Volga Dnepr.

Silk Way Airlines' fleet included this An-12BK wearing the new-style registration 4K-AZ23 (originally applied in error as 4KAZ-23; ex-11715, c/n 8345605). The aircraft has been demilitarised by removing the tail turret and overpainting the tail gunner's station glazing. Its subsequent fate is unknown. *Yefim Gordon archive*

Turan Air

Founded in 1994 as an Azerbaijan-US joint venture, Baku-based Turan Air [3T/URN] had a fleet consisting of four Tu-154B-2s (re-exports from Hungary and Latvia) and two ex-Polish Tu-154Ms. These included, left, 4K-85524 (ex-Latavio YL-LAG, c/n 81A524) named *Baki* (Baku in the Azeri language) in basic Aeroflot colours and, below, 4K-474 (ex-Malev Hungarian Airlines HA-LCP, c/n 81A474) in basic Malev livery. Currently two Tu-154B-2s remain on the list and have been augmented by a Boeing 727-251 Advanced. *Yuriy Kirsanov*

Belgium

TNT

Based at Liege, TNT Airways [3V/TAY] – a subsidiary of the TNT Global Express, Logistics & Mail company – operates assorted freighters, mostly registered in Belgium. In 2002, however, the fleet was augmented by an Egyptian-registered Tu-204-120C, SU-EAJ, leased from Sirocco Leasing. These photos show the airline's second Tu-204-120C, SU-EAK (c/n 1450743164033), awaiting completion at the Aviastar factory in Ul'yanovsk in 2006. *Marina Lystseva*

Belarus

The Belorussian flag carrier **Belavia [B2/BRU]** was formed in 1992 as the successor of Aeroflot's Belorussian CAD, with divisions in Minsk (two), Brest, Gomel' and Mogilyov. Until 1993 the carrier operated under the trading name Belarus' and the Aeroflot flight code. The original fleet included18 An-24B/RVs, seven An-26s, 19 Tu-134A/AKs, 17 Tu-154B-1/-2s, seven Tu-154Ms and eight Yak-40s. In 1996 the airline was downsized dramatically when its main branches became separate enterprises (Gomelavia, Mogilyovavia and Minskavia, the latter operating from Minsk-1 (Loshitsa).

The present-day Belavia based at Minsk-2 (Velikiy Dvor) operates scheduled and charter services with two Boeing 737-500s, two An-24RVs, two Tu-134As, four Tu-154Ms and one VIP-configured Yak-40 'Salon'. The Tupolev twinjets are due for replacement with Bombardier CRJ700s. The airline's scheduled services include Berlin, Frankfurt/Main, Istanbul, Kaliningrad, Kiev, Larnaca, London, Moscow, Paris, Rome, Stockholm, Tbilis, Tel Aviv, Warsaw and Yerevan.

Belavia

One of Belavia's An-24RVs based at Minsk-1, EW-47291 (c/n 07306601), approaches Moscow-Sheremet'yevo. Note the post-1997 red/green/white Belorussian flag (similar to the flag of the former Belorussian SSR) introduced when Russia and Belarus formed a new union. *Sergey Sergeyev*

Showing some signs of weathering, Tu-154M EW-85748 (c/n 92A924) departs Prague-Ruzyne. All of the airline's Tu-154s retained the Aeroflot cheatline. Note the position of the flag. *Martin Novak*

In the early days of the CIS, some Belorussian airliners, including Tu-134A EW-65803 (c/n 3352103) seen taking off from Moscow-Sheremet'yevo in 1994, wore this livery with Cyrillic Belarus' titles and the old-style flag on a blue band on the tail. Originally they even wore the old CCCP- prefix and small Aeroflot titles. EW-65803 gained the Belavia livery in 1995, only to be retired a year later. *Sergey Sergeyev*

Immaculate-looking Tu-134A EW-85145 (c/n [93]60985, f/n 4910) sits on a rain-soaked apron at Minsk-2. This aircraft has Belavia's own cheatline with pinstripes above and below. *via Peter Davison*

Executive-configured Yakovlev Yak-40 'Salon 2nd Class' EW-88161 (c/n 9611546) is seen at Minsk-2 on the same day; the white plug in the foremost cabin window shows where a coat closet is located on this version. The aircraft is operated by Belavia's Mogilyov division. *via Peter Davison*

Gomelavia

Established in 1996, Gomelavia [YD/GOM] is the former Gomel' division of Belavia. Up to 2004 it operated jet airliners (three Tu-134As, two Tu-134AKs, four leased Tu-154B-2s and two Tu-154Ms). These and the An-26 freighters have now been retired, returned or sold, leaving three An-24RVs as the carrier's only passenger aircraft; the Mil' Mi-2 utility helicopters have also been disposed of. An-24RV EW-46483 (c/n 27308101) shown here is no longer in the fleet, having been transferred to Belavia in 1995; note the registration carried entirely under one wing. The ones still in service have all-white fuselages. However, Gomelavia continues cargo operations with a single An-12BP and six IL-76TDs, which are new additions to the fleet.
via Viktor Kravchenko

TransAviaExport

TransAviaExport Cargo Airlines [AL/TXC] were founded in 1992 with a fleet of 24 IL-76MDs as the commercial division of the Belorussian Air Force operating from Machoolishchi AB near Minsk. Here, demilitarised IL-76TD 'Falsie' (actually an IL-76MD – note tail gunner's station with overpainted windows) EW-78843 (c/n 1003403082, f/n 7801) climbs away, showing the airline's current colours with a small post-1997 flag on the nose.
Yefim Gordon archive

Belorussian Government

Smartly painted Tu-154M 'Salon' EW-85815 (c/n 95A1010), delivered in 1998 with a 49-seat interior, is operated by the Belorussian government flight and is seen here visiting Moscow. Yet now it has ceded its 'No 1 Presidential jet' role to a Boeing BBJ. *Yuriy Kirsanov*

For shorter trips the Belorussian President Aleksandr G Lukashenko uses this 16-seat Yak-40 'Salon 1st Class', EW-88187 (c/n 9620748), as well as at least two Mil' Mi-172 VIP helicopters wearing the same basic livery. *Yefim Gordon archive*

Bolivia

AeroSur Express

Based at Santa Cruz-Viru Viru airport, the private airline AeroSur [5L/RSU] ('South Airlines' in Spanish) leased Yak-40K OB-1653 (c/n 9041860) from the Peruvian carrier T doble A (Transportes Aereos Andinos) in September 1998; by September 2003 it had been placed in storage. Earlier, in 1995-1997, the same aircraft had been leased from another Peruvian airline (Expresso Aereo) under its then-current registration OB-1606. *via Martin Štepanek*

Bulgaria

Air Sofia

The cargo airline Air Sofia [CT/SFB] was established in 1992. The mainstay of the fleet is the good old An-12, of which 15 were operated over the years, including demilitarised An-12BP LZ-SFS (ex-Polish Air Force '51 Red', c/n 6344308). The fleet also includes a single An-26; single examples of the An-24 and even the An-22 were operated earlier. Most of the aircraft are white overall; a curious feature of Air Sofia's An-12s is that they lack the usual radar in a chin radome. Air Sofia often leases its An-12s to other carriers. *Martin Novak*

Balkan Bulgarian Airlines

Established in 1949 when the country came under Soviet influence, the Bulgarian/Soviet enterprise **TABSO – Bulgarian Air Transport** was the nation's first flag carrier. It started out with the piston-engined Lisunov Li-2 (DC-3 derivative) and IL-14, these were followed by the IL-18 turboprop and the Tu-134, which was the first jet airliner to see service in Bulgaria. In 1968 TABSO sprouted a division named Bulair which performed charter flights with IL-18s and utility work with An-2s. That same year the flag carrier was renamed **Balkan Bulgarian Airlines [LZ/LAZ]**, re-absorbing Bulair in 1971.

Balkan continued operating the An-24B/RV (16), IL-18V/E (10), Tu-134/Tu-134A (13), Tu-154/Tu-154A/B/B-1/B-2/M (47, including leased ones), and Yak-40 (12). Cargo operations were performed by IL-18s and An-12s (8). Mi-2 and Mi-8 helicopters were also operated on utility and passenger flights.

After the fall of Todor Zhivkov's communist regime Balkan started a major fleet renewal programme, no longer relying on Soviet hardware. The change was heralded by the arrival of three leased Boeing 737-500s in 1990 to replace the Tu-134As, which were withdrawn in 1995. The 737s were followed by Airbus A320s (as a Tu-154 replacement) and long-haul Boeing 767-200ERs in 1991-92, while the An-24s gave place to ATR42 turboprops. The Yak-40s and the remainder of the Tu-134 and Tu-154 fleets were transferred or sold to Hemus Air from 1986 onwards. However, Balkan still leased Tu-154Ms from Russian carriers during the summer charter season until 2001.

At the end of the 1990s Balkan started suffering serious financial troubles. An attempt was made to rescue the airline by selling it to the Israeli financial company Ze'evi Group which also owns the Israeli airline Arkia. The new shareholder took up the issue and started implementing a fleet renewal programme. This change of ownership had an immediate adverse effect – some Middle Eastern nations whose cities had, until then, been served by Balkan denied access to Balkan aircraft because the carrier had 'sold out to the enemy'! However, the Israeli investment came too late and in December 2002 Balkan finally ceased operations, ceding the flag carrier role to a new outfit called Bulgaria Air.

'Glass-nosed' Tu-134A LZ-TUK (c/n 1351209), seen here in its original 1970s livery, was originally a VIP jet operated for the Bulgarian government. It was later refitted as a 76-seater and transferred to Balkan, plying the Sofia-Prague route, among other things. Note the 16 cabin windows to port characteristic of Tu-134s in the first three production batches (10-12); this was reduced to 15 on later aircraft. *Martin Novak*

Pictured on short finals to Moscow-Sheremet'yevo, LZ-TUM (c/n 3351906) in 1980s Balkan colours was the first Tu-134AK VIP jet to be built in the 'radar-nosed' export version. It was sold to Russian charter carrier ShaNS-Air in 1995 as RA-65940. *Yuriy Kirsanov*

Seen here visiting Prague-Ruzyne, IL-18V LZ-BEU (ex-Yemen Airways 4W-ABO, c/n 183005905) in Balkan's 1980s colours emphasises the ability of a modern livery to give a vintage airliner a completely new look. It was sold to Bulgarian Flying Cargo in 2000 as LZ-BFU. *Martin Novak*

A fine landing study of Tu-154B-2 LZ-BTS (c/n 80A422). This aircraft was sold to Aerotrans of Kazakhstan in 2000 as UN-85422, retaining basic Balkan colours.
Yuriy Kirsanov

As it started phasing out its own Tu-154s, Balkan was forced to lease them from other carriers as required. Tu-154-100 OM-AAB (c/n 98A1015) was leased from Slovakian Airlines from October 2001 to September 2002 in the lessor's livery with Balkan titles.
Sergey and Dmitriy Komissarov collection

Balkan also flew cargo services with a small fleet of civil-configured An-12s. Voronezh-built LZ-BAF *River of Maritza* (c/n 402408) wears the standard livery.

Tashkent-built LZ-BAC (c/n 6343708; later LZ-CBE with Heli-Air) has a non-standard scheme minus the green colour.
Both Yefim Gordon archive

Balkan Holidays

A charter airline called Balkan Holidays or BH Air [–/BGH] leased three Tu-154Ms, including LZ-HMQ (c/n 86A743), in May 2002, repainting them in this attractive multi-coloured livery. *Sergey and Dmitriy Komissarov collection*

Inter Trans Air

The cargo charter airline Inter Trans Air [–/ITT] operated four An-12s. The first of these, demilitarised An-12B LZ-ITA (c/n 3341004), is seen here in its better days; note the support under the tail. The airline went out of business in 2004; this aircraft was withdrawn from use at Gorna Orekhovitsa, Bulgaria. *Yefim Gordon archive*

Bulgarian Air Charter

Founded in 2000, Bulgarian Air Charter [–/BUC] operated eleven Tu-154Ms in a green/white livery over the years, mostly leasing them from Russia and Azerbaijan as required. LZ-LCX is an exception, being ex-Balkan LZ-BTX (c/n 86A744). *Sergey and Dmitriy Komissarov collection*

Heli Air

Contrary to its name, Heli Air Services [–/HLR] of Sofia, founded in1990, has more fixed-wing aircraft than helicopters! It operates both passenger/utility and cargo services; the freighter fleet consists of four An-12s, including Tashkent-built An-12A LZ-CBG (ex-Aerostatus RA-11370, c/n 2340804) – one of the oldest active examples as of now. *Eero Saarela*

Hemus Air

Hemus Air [DU/HMS] started operations in 1986 as a department of Balkan Bulgarian Airlines providing ambulance services, flight calibration and aerial photography. In 1991 it became a separate legal entity and was named Hemus Air (in antiquity the Balkan Mountains were known as Haemus Mons). The company was privatised by Bulgarian corporate investors in 2002.
Hemus Air operates both scheduled and charter services. The fleet included eight Tu-134A/AK/Bs, nine Tu-154Ms and seven Yak-40s; yet the Tupolev twinjets and most of the trijets have been disposed of, giving way to BAe 146-200s and a Boeing 737-500.

Tu-134A-1 LZ-TUH (c/n 7360142, f/n 4305) was acquired in 1997 from CSA, which had operated it as OK-HFM. After flying in this *very* restrained livery for nearly ten years it was sold to Russia. *Sergey Sergeyev*

Yak-40 LZ-DOF (c/n 9521541) illustrates the original multi-coloured livery of Hemus Air with Bulgarian titles. Note the small Yamaha titles and the obscure badge on the nose; since this is hardly the aircraft's name, it may be surmised that LZ-DOF was chartered by Yamaha. This 30-seat aircraft is one of three Yak-40s currently on strength. *via Bob Archer*

Hemus Air was an active operator of the Tu-154M at one time. Most of the aircraft were former Balkan machines; however, LZ-HMP (c/n 86A733) seen here visiting the UK was the former 4K-733 bought in 2000 from Turan Air, retaining the latter's basic livery and a modified version of the Turan Air logo. In 2002 it was sold to Bulgarian Air Charter as LZ-LCV. *via Bob Archer*

VIA

The charter carrier **VIA – Varna International Airlines [VL/VIM]** was formed in mid-1990, flying a fleet of six Tu-154Ms. VIA went on to become one of Bulgaria's better inclusive-tour airlines. In addition to their own services, the airline leased the trijets to other carriers. However, in late 2006 the carrier phased out the Tu-154M because of noise restrictions and high operating costs. Currently VIA operates three Airbus Industrie A320-232s.

The first of the six Tu-154Ms operated by VIA – Varna International Airlines, LZ-MIG (c/n 90A840), is pictured here in its original livery with small titles and small tail logo on a white background. *Yefim Gordon archive*

In a pun on the abbreviated version of the name, the trijets soon received much bolder titles reading 'VIA est vita' and a larger tail logo. LZ-MIG illustrates the new livery during one of its visits to Prague-Ruzyne. *Martin Novak*

Vega Airlines

Founded in 1998 and based in Plovdiv, the cargo carrier Vega Airlines [–/VEA] has a fleet of six An-12s sourced from such far-apart places as Bulgaria, Russia, Uzbekistan and Angola. All aircraft have proper names; demilitarised An-12BP LZ-VEC (ex-UK 93920, c/n 6344810) is named *Tsar Simeon Veliki* (Czar Simon the Great, Bulgaria's ruler in the late 9th/early 10th century). *Yefim Gordon archive*

Central African Republic

Centrafican Airlines

Founded in 1998, Centrafrican Airlines [GC/CET] took over almost the entire fleet of the defunct Air Pass, nominally a Swazi company. Indeed, it is very probable that both airlines had a common owner and the 'acquisition' was just a change of name. Besides Bangui (the capital of the Central African Republic), the carrier was based at Sharjah. Its sole Tu-154M, TL-ACF (formerly Air Pass 3D-RTP, c/n 91A895) is seen here landing at the latter location in 1999. In 2000 the aircraft was sold to Moldtransavia as ER-TAG.
Peter Davison

China

CAAC

The **Civil Aviation Administration of China (CAAC) [CA/CCA]** emerged in 1955 as a one-size-fits-all 'Chinese Aeroflot' tasked with passenger/cargo transportation and various utility work. In a similar manner to Aeroflot's organisation, it was divided into territorial directorates with offices in Beijing, Shanghai, Guangzhou, Chengdu, Lanzhou (later moved to Xi'an), Shenyang and Urumqi.

Over the years the airline operated such Soviet types as the Li-2, An-2, IL-14 and An-24, as well as the Mi-8 helicopter (50+), on the domestic network, the IL-18 (up to 17) and IL-62 (six) on international flights, and the An-12 (two) on cargo services. Licence-built versions of the Antonov types were also flown. Due to the deterioration of Sino-Soviet relations in the 1960s China started looking to the West and increasingly relying on US-built, European and indigenous types for civil air transportation. Yet this did not stop China from buying further Soviet airliners, such as the Tu-154M.

In 1987, deregulation set in; CAAC's aviation regulation division and the airline operation division were separated, and the latter's directorates became new airlines – Air China (which inheritec the flight codes of the original CAAC), China Eastern Airlines, China Southern Airlines, China Northwest Airlines, China Northern Airlines and China Southwest Airlines, each named after their main operation areas. Further deliveries were made to these and other airlines.

Illustrating CAAC's early days, IL-18V B-218 (originally 218; c/n 184007601) is seen here undergoing routine maintenance; note the early-style strake aerial on top of the fuselage. This aircraft is now preserved in Lanzhou.
Yefim Gordon archive

CAAC's fifth Tu-154M, B-2605 (c/n 85A725) taxies at Beijing-International. After the deregulation of CAAC this aircraft, which was assigned to the Xian Directorate, served on with China Northwest Airlines. *Peter Davison*

Air China

Air China [CA/CCA], which started operations on 1st July 1988, is the successor of CAAC's Beijing directorate. Although the fleet consisted mostly of Boeing 737s, 747s and 767s, in the 1990s Air China operated one Tu-154M and two An-12BPs, as well as several Xian Y7s (licence-built An-24s). These have now been withdrawn, but in 2005 Russia made a comeback when Air China ordered three new Tu-204-120C freighters due for delivery in 2006 and 2007.

One of Air China's two Soviet-built An-12BPs, B-3152 (formerly B-203, c/n 8345303). The aircraft was retired by 2000. *China Aircraft*

In 2004 Tupolev made a comeback on the Chinese civil aviation market, securing orders for five Rolls-Royce-powered Tu-204-120CE freighters. B-2871 (c/n 1450743664030) is the first of the three ordered by Air China – and the first Tupolev aircraft to wear this carrier's colours. *Tupolev PLC*

The future B-2871 flies over its birthplace, the massive Aviastar plant, with the apartment blocks of Ul'yanovsk in the background. The aircraft is still in primer finish with the last two digits of the c/n on the tail. The inscriptions read 'Certification tests', 'Sirocco-KNR' (*Kitayskaya narodnayarespooblika* – People's Republic of China) and 'Tu-204-120SE' (as this version is known in Russia). *Tupolev PLC*

B-2871 touches down after a pre-delivery test flight. Note the still-unpainted service door. *Tupolev PLC*

China General Aviation

China General Aviation Corp. [–/CTH] was established in 1989, performing mostly utillity/survey work with fixed-wing aircraft and helicopters of various origin. However, it also flew scheduled serviced with Xian Y7-100Cs, Boeing 737-300s and eight Yak-42Ds, including B-2752 (c/n 4520424116664, f/n 0414). When one of them crashed, the rest were retired and sold; B-2752 became YL-LBV with Concors. In January 1998 CGAC merged into China Eastern Airlines. *via Bob Archer*

China Southwest Airlines

The fleet of China Southwest Airlines [SZ/CXN] based in Chengdu, Sichuan Province, included five Tu-154Ms. When B-2622 crashed in 1999, the surviving examples were grounded and eventually sold to Russia; B-2617 (c/n 88A791) became RA-85136 with Blagoveshchensk Airlines in 2001. *Yefim Gordon archive*

China United Airlines

China United Airlines [CN/CUA] were established in 1986 as the commercial division of the People's Liberation Army Air Force (PLAAF) with bases at Beijing-Nan Yuan, Beijing-Fo Shan AB and Nanjing. It operated scheduled and charter services in co-operation with local enterprises. The fleet included quasi-civil IL-18s (four), Tu-154Ms (16) and IL-76MD 'Falsies' (20), as well as Boeing 737-300s and Canadair Challenger 601s. Other aircraft reportedly operated by CUA had military serials. Some of the carrier's aircraft were airborne command posts or electronic intelligence platforms.

In November 2002 China United Airlines terminated its scheduled service across the country. In 2003 the airline was liquidated altogether due to a new regulation imposed by the Chinese government which forbids the People's Liberation Army's involvement in commercial activities. However, on 4th June 2005 the Civil Aviation Administration of China gave the approval for China United Airlines to be reconstructed by Shanghai Airlines and China Aviation Supplies (CASGC). The reborn airline still operates from military airfields but no longer flies Russian types.

IL-76MD 'Falsie' B-4032 (c/n 1013409289, f/n 8303), is one of the initial batch of three delivered in 1991. *Yefim Gordon archive*

The red cheatline and tail of Tu-154M B-4023 (c/n 88A770) reveal its origin: the aircraft was bought in September 1992 from ČSA Czechoslovak Airlines with whom it had flown as OK-TCB. *via Martin Štepanek*

China Xinjiang Airlines

China Xinjiang Airlines [XO/CXJ] based in Urumqi, the capital of the Sinkiang-Uigur Province, was founded in 1985 as the successor of CAAC's Urumqi Directorate. The airline had a secondary hub in Changzhou near Nanjing, Jiangsu Province. It operated a mix of Russian and Western aircraft. The former included eight Tu-154Ms, four of which were inherited from CAAC and two were leased from Russia. In 1993 China Xinjiang Airlines became the first and only export customer for the IL-86 wide-body airliner, purchasing three examples.

In 2001 the Tupolev trijets were sold to various CIS airlines. The IL-86s stayed until 2003 when the carrier was taken over by China Southern Airlines.

Tu-154M B-2603 (c/n 85A718) caught by the camera on short finals to Moscow-Sheremet'yevo illustrates China Xinjiang Airlines' late blue-tail livery introduced by 1998. In April 2001 the aircraft was sold to Imair as 4K-AZ17. *Yuriy Kirsanov*

B-2019 (c/n 54383210100), the last of the three IL-86s delivered, is seen on finals to Moscow-Sheremet'yevo in China Xinjiang Airlines' early white-tail livery. It received the late blue-tailed version before being sold to Kras Air as RA-86926. *via Viktor Kravchenko*

Sichuan Airlines

Sichuan Airlines [3U/CSC] based in Chengdu, Sichuan Province, was founded in 1986. In 1991-1993 the carrier took delivery of five new Tu-154Ms which supplemented the Xian Y7-100s used on short-haul services. Apart from one example which had been sold to China Xinjiang Airlines, the trijets returned to Russia in 2001. In August 2002 Sichuan Airlines was reorganised and now operates an all-Western fleet.

Tu-154M B-2626 (c/n 91A894) shows the attractive livery of Sichuan Airlines. In December 2001 it was sold to the Pulkovo Air Enterprise, becoming RA-85185 No 2. Tragically, this aircraft crashed near Donetsk, the Ukraine, on 22nd August 2006, killing all on board. *Yefim Gordon archive*

Colombia

SELVA

Founded in 1978, **Servicios Aereos del Vaupes** based at Villavicencio, Vaupes Province, and doing business as **SELVA [–/SDV]**, was one of several Colombian operators of the An-32. Eight examples were used at various times for carrying anything from newspapers to convicts. Unfortunately four of the aircraft were lost in accidents

Bought in 1994, An-32B HK-3930X (formerly RA-48015 with Stela airlines of Orenburg, c/n 3309) was one of several operated by SELVA Colombia. It is shown here in early colours. The aircraft crashed on landing at Rionegro, Colombia, in thick fog on 22nd December 1998. *Yefim Gordon archive*

Congo-Brazzaville

Aero-Service

Aero-Service [BF/RSR] **was established in 1967, flying scheduled and charter passenger and cargo services from Brazzaville. The airline has operated such Russian types as the An-12, An-24 and Yak-42, leasing them as required. Here, An-24RV ER-AZB (c/n 27307507) leased from the Moldovan carrier Pecotox Air in 2003-2004 displays the vivid 'wildlife' colours of Aero-Service; note the buffalo motif on the tail.** *Yefim Gordon archive*

Cuba

Cubana

The airline was established on 8th October 1929 as Compañía Nacional Cubana de Aviación Curtiss. Pan American Airways System bought Cubana in 1932, and the word Curtiss was dropped from the airline's name. In 1944 the name was changed to Compañía Cubana de Aviación SA. Pan American sold a majority share in the airline to Cuban investors in 1945.

After Fidel Castro came into power in 1959, Cubana was nationalised, and three private Cuban air carriers were merged into Cubana. This effectively consolidated Cuban commercial aviation under government ownership. Cubana's name was changed to Empresa Consolidada Cubana de Aviación SA. Until early 1961 the carrier had operated American and British types, but a US embargo forced it to turn to the Soviet Union for new hardware; the first type delivered was the IL-14 in 1961 (14 delivered), subsequently replaced on local routes by the An-24 (since 1966) and Yak-40 (since 1975). From 1964 the IL-18 (eight), followed in 1967 by the IL-62 (one leased) and in 1977 by the IL-62M (15), were used on the longer routes to Canada, the Caribbean, South America and Europe. The Tu-154 (nine) and Yak-42 trijets were introduced in 1980 and 1990 respectively on medium-haul services. The An-12, An-26 and IL-76MD (the latter owned by the Cuban Air Force) were used on cargo services. As a result, Cubana became the Latin American carrier with most experience in the operation of Soviet-built aircraft. However, Cubana's old Bristol Britannias and leased Douglas DC-8s were still used in the 1970s.

By early 1980, Cubana flew scheduled services to Sub-Saharan Africa, the Middle East and North Africa, Eastern and Western Europe, Canada and various Latin American and Caribbean destinations.

After the demise of the Soviet Union in 1991, Cubana was given license to buy or lease aircraft from elsewhere. At the same time the carrier faced the formidable challenge of improving its service standards, which were far below pre-Castro levels. The airline's reputation was marred by a spate of fatal accidents in the 1990s, and, with Russia moving towards a market economy, spares for Cubana's Soviet-built aircraft became hard to find. Cubana's management made efforts to renovate its fleet, ordering the IL-96-300 long-haul wide-body and the Tu-204-100 medium-haul narrowbody from Russia in 2005.

The current fleet includes six An-24RVs, six Yak-42Ds, four IL-62Ms and two IL-96-300s (with two more on order). Two Tu-204-100s and one Tu-204-100C freighter are on order. Some of Cubana's aircraft are VIP jets used by Cuban government officials on trips abroad. The airline's main base is Havana-José Martí.

An-24RV CU-T1267 (c/n 47309907) seen here staging through Prague-Ruzyne in 1987 on delivery to Cuba in the early colours of Cubana was a second-hand aircraft, being previously operated by Aeroflot as CCCP-46696. Note that the registration has been mispainted as 'CUT-1267'. *Martin Novak*

Being short of capacity, Cubana was forced to lease aircraft from other socialist bloc countries occasionally. IL-18V YR-IMG (c/n 184007301) was leased from TAROM in February-October 1990, wearing this hybrid livery combining the TAROM cheatline with Cubana tail colours and titles. *Martin Novak*

In due course the IL-62M replaced the IL-18 on the Cuban flag carrier's longest routes. Here, CU-T1283 (c/n 4053823) is seen arriving at Helsinki-Vantaa. *Eero Saarela*

In the 1990s Cubana introduced a new livery with a white fuselage and much bolder titles, illustrated by IL-62M CU-T1217 (c/n 3933232) on approach to Moscow-Sheremet'yevo.

Tu-154M CU-T1264 (c/n 85A720) is seen staging through Prague-Ruzyne in old colours. This aircraft was lost in a fatal crash at Quito-Mariscal Sucre International on 29th August 1998. *Martin Novak*

Cubana's Yak-42Ds wore a version of the old livery with white undersides instead of the usual grey colour. Here CU-T1279 (c/n 4520424914057, f/n 0210) arrives at Moscow-Bykovo for refurbishment at Aircraft Repair Plant No 402. *Yuriy Kirsanov*

Cubana's Yak-40s also sported a somewhat unusual livery with a red centre engine intake trunk. They were used both as feederliners and as executive jets operated for the Cuban government. CU-T1221 (c/n 9841159) was one of the regular airline-configured examples; it was sold to competing Aerocaribbean by 1993. *Martin Novak*

Cubana operated both specialised freighters and airliners converted to freighters. The former types included two IL-76MD 'Falsies' (with 'IL-76MD' nose titles but no tail gunner's station) owned by the Cuban Air Force (DAAFAR) but painted in Cubana colours and occasionally used by the airline. The first of the two, CU-T1258 (c/n 0043454615, f/n 4104), is depicted here. It was reregistered CU-C1258 in 1996 (C for *Carga*), then became CU-C1419 in 2000 and was eventually sold to the Moldovan airline Aerocom as ER-IBE in 2002. *via Yefim Gordon*

Cubana's first IL-96-300, CU-T1250 (c/n 74393202015), lands on runway 30 at Zhukovskiy, arriving for the MAKS-2005 airshow. *Yefim Gordon*

Aerocaribbean

Aerocaribbean [7L/CRN] based in Havana started operations on 2nd December 1982, operating domestic flights and regional charters to supplement the flag carrier Cubana. Its fleet consists of aircraft transferred from Cubana and acquired elsewhere. One An-24RV, two An-26s, three IL-18s and three Yak-40s are currently on strength.

Aerocaribbean's aircraft wore a variety of liveries. An-26 CU-T111 (formerly CU-T1230 with Cubana, ex-CCCP-47333, c/n 7207) had this smart version with an orange cheatline. It is no longer in service. *Martin Novak*

Aerogaviota

Aerogaviota [KG/GTV] (*gaviota* is Spanish for 'seagull') was formed in 1994 as the commercial division of the DAAFAR, flying mostly An-26 freighters and Mil' Mi-8/Mi-17 helicopters from Havana-Jose Marti. The current fleet includes 11 An-26s *sans suffixe*, seven An-26Bs, one Mi-8T utility chopper, three Mi-8P heli-liners and two Mi-17s, as well as four Aerospatiale/ Alenia ATR 42-500 regional turboprops introduced in 2000. Here, Aerogaviota's An-26 CU-T1435 (ex-CCCP-47339, c/n 7702) is seen visiting Moscow.
Yuriy Kirsanov

Czechoslovakia / Czech Republic

ČSA Czech Airlines

The nation's flag carrier ČSA České Aérolinie (Czech Airlines) [OK/CSA] is an air carrier with a long history. It started life on 19th October 1923 as ČSA Československé Statni Aérolinie (Czechoslovak State Airlines), operating Aero A-14 and A-10 biplanes. These were augmented in the pre-war years by such types as the Junkers F13, Farman F-60 Goliath, Letov S-32, Aero A-23, Fokker F-VIIb-3m and Ford 5-AT Trimotor. Even the Saunders Roe A.19 Cloud amphibian was operated on routes to destinations on the Adriatic Sea.

Then came the first break-up of Czechoslovakia on 14th March 1939, followed the next day by German occupation of Czechia and Slovakia. On 17th August that year, shortly before the outbreak of World War Two, the airline (now called Czech State Airlines) went into liquidation.

After the war and the re-formation of the Czechoslovak Republic (the so-called Second Republic), the airline was reborn as the nationalised ČSA Československé Aérolinie (Czechoslovak Airlines), the sole air carrier in the republic, pursuant to government decrees dated 14th September and 24th October 1945. Initially the Douglas DC-3 made up the backbone of the fleet and plans were in hand to buy the DC-6 and other Western types. However, on 25th February 1948 Czechoslovakia came completely under Soviet influence and these plans had to be abandoned due to an embargo on Western hardware deliveries. From then on for many years ČSA operated only Soviet or indigenous types. In the late 1940s/early 1950s the carrier's international services were curtailed dramatically and traffic volumes declined. The airline tried to offset this by expanding its domestic route network.

The first Soviet-supplied equipment came in 1949 when the first of ten IL-12B twin-engined airliners was delivered. These were followed by eight Li-2Ps in 1952. Five years later ČSA started taking delivery of the Avia-14 – a licence-built version of the IL-14P (29 were operated until 1977). On 15th June 1956 ČSA entered the jet age, receiving the first of six Tu-104A medium-haul airliners. From 1960 onwards these were augmented by 12 IL-18 turboprops which helped rebuild the carrier's route network, the type staying in service until 1984. Later the long-haul routes were taken over by nine IL-62s and six IL-62Ms operated since 1969 and 1979 respectively. Short/medium-haul services were performed by three Tu-124Vs delivered in 1964-65, later replaced by 14 Tu-134As from 1971 onwards, and by 17 Yak-40 feederjets. Unlike many Soviet bloc airlines, ČSA never had any Antonov aircraft.

Seven Tu-154Ms were delivered in 1988-90. In 1989 Czechoslovakia became the Czecho-Slovak Federal Republic; on 1st January 1993 the latter was divided into two independent states and the airline was renamed ČSA Czech Airlines.

However, the passengers were distrustful of Russian aircraft, whose operating economics left a lot to be desired into the bargain. Hence in 1991 the carrier started re-equipping with Airbuses, Boeings and ATRs. On 10th January 2000 Tu-154M OK-UCE flew a farewell service, ending ČSA's involvement with Russian aircraft.

ČSA's fourth IL-62 *sans suffixe*, OK-ABD *Košice* (c/n 10902) in the original version of the 1971 livery (with 'OK jet' tail titles) shares the apron at Prague-Ruzyne with Tu-134A-1 OK-HFM, IL-18V OK-VAF and a further Tu-134A. *Martin Novak*

IL-62M OK-JBI *Plzeň* (c/n 2932748) displays an intermediate livery tried in the early 1990s in order to give the aircraft more appeal. Note the 'Czechoslovak Airlines' titles (not 'Československé Aérolinie'). This aircraft was sold consecutively to several Czech air carriers and eventually to Russia as RA-86597. *Martin Novak*

Opposite page: A picture from days gone by: IL-18V OK-PAI (formerly IL-18V 'Salon' OK-BYP with the Government flight, c/n 181003105) taxies at Prague-Ruzyne in pre-1971 colours with what was known locally as *fousy* (moustache) on the nose. The aircraft is now preserved at the aviation museum in Sinsheim, Germany. *Martin Novak*

IL-62M OK-OBL *Ostrava* (c/n 4445032) was painted in another version of the 1990s livery with a red 'OK' logo on a white tail. Note that the radome does not have a black tip, contrary to usual practice. The aircraft was sold to Yana Airlines of Kampuchea in 1999 as XU-229. *Martin Novak*

The airline's sixth Tu-134A, OK-EFK (c/n 4323130, f/n 2608) in the type's original colours with 'OK jet' tail titles. It was retired on 30th January 1995 and broken up at Prague-Ruzyne ten months later. All of ČSA's Tu-134As were radar-nosed; the black paint on the radome was a local addition. *Martin Novak*

Another ČSA Tu-134A, OK-HFL (c/n 7349913, f/n 4105) in the ultimate colours with 'Czech Airlines' titles and tail 'pennants' is seen visiting Moscow-Sheremet'yevo. It was sold to Hemus Air in 1997 as LZ-TUJ. *Yuriy Kirsanov*

ČSA's fleet included 17 Yak-40 feederliners. The last example delivered, Yak-40K OK-HER (c/n 9741156), rests between flights at Prague-Ruzyne in the ultimate livery worn by the type. It was sold to Belorussia in 1992 but never operated there, becoming UR-87590 with UES-Avia in 1997. *via Yefim Gordon*

ČSA's second Tu-154M, OK-TCB *Karlovy Vary* (c/n 88A770), wore the late 1971-standard livery (with 'OK csa' tail titles) throughout – that is, until sold to China United Airlines in September 1992 as B-4023. Note the city crest of Karlovy Vary aft of the rear entry door. *Martin Novak*

Tu-154M OK-VCG *Luhačovice* (c/n 90A838) wore this interim white-tailed livery with Czechoslovak Airlines titles for a while. It was sold to Russia in 2000 and assigned the registration RA-85089 No 2 but never wore it, being leased to Kish Air straight away as EP-LBR. *Martin Novak*

OK-TCD *Trenčianské Teplice* (c/n 88A792) in current ČSA colours is caught by the camera a few seconds before landing on runway 07R at Moscow-Sheremet'yevo. Since January 2000 this aircraft is in service with Bashkirian Airlines as RA-85847. *Dmitriy Petrochenko*

Škoda Air

Škoda Air [–/SOA], a subsidiary of the Skoda industrial concern based in Plzeň, leased a single Yak-42D, RA-42411 (c/n 4520422219043, f/n 0315), from Bykovo-Avia in June 1994, operating it until mid-1995. Here the aircraft also carries the titles of Mostarez Air, a Czech charter airline, and the oval logo of the Miky Tour travel agency for which it was operated. The tail logo is that of the previous lessor, the Bosnian carrier Vardar Air. *Yefim Gordon archive*

Travel Service Airlines

Prague-based Travel Service Airlines [–TVS] operated this smartly painted Tu-154M, OK-VCP (c/n 90A858), in 1997-99, acquiring it from the defunct Ensor Air. The aircraft was eventually sold to Omskavia as RA-85841, departing to Russia on 17th October 1999.
Yefim Gordon archive

Espe Air Prague

Short-lived charter carrier Espe Air Prague (actually SP Air) bought IL-62 *sans suffixe* OK-GBH *Usti nad Labem* (c/n 62404) from ČSA in 1993. A year later the aircraft was sold to Georgia Air Prague, serving with this carrier until its retirement – whereupon it became a restaurant in Germany.
Martin Novak

Ensor Air

Ensor Air [E9/ENR], likewise based in Prague, operated two former VIP jets bought from the Czech government flight. One was Tu-154B-2 OK-LCP (formerly Tu-154B-2 'Salon' OK-BYB, c/n 81A488) bought in 1992; it had a red/blue cheatline. This aircraft was sold to Aeroflot in January 1993 as RA-85488.

The carrier's other aircraft was this Tu-154M, OK-VCP (ex-Tu-154M 'Salon' OK-BYP) bought in 1993, which wore an altogether different and more modern colour scheme. The trijets were jointly operated with the Excel travel agency, carrying its logo on the centre air intake trunk or, in this case, on the engine nacelles. In 1997 OK-VCP was sold to Travel Service Airlines (see opposite page). *Martin Novak*

Czech Government

The Czechoslovak Federal Government Flight (LOMV – *Létecky oddíl ministerstva vnitra*) and its Czech successor, the State Flight Unit (SLU – *Státní létecky útvar*) operated VIP versions of various Soviet types since the early 1950s, including the IL-14, IL-18, IL-62/IL-62M, Tu-134A, Tu-154B-2/Tu-154M and Yak-40. OK-BYJ, a convertible Yak-40K (c/n 9821257), is shown here in its final civilian colours; it is now Czech Air Force '1257'. *via Martin Štepanek*

Djibouti

Daallo Airlines

Daallo Airlines [D3/DAO] started operations on 20th March 1991 with one Cessna aircraft. Soviet aircraft (the Tu-154, IL-18 and An-24) leased from Tajik and Kyrghyz carriers were largely used in the 1990s, along with Boeings and Airbuses. The airline operates from Djibouti and Dubai, UAE.

Seen at Sharjah, An-24RV EY-47693 (c/n 27307510) was leased thrice from Tajikistan Airlines in 1995, 1999 and 2000. *Yuriy Kirsanov*

Tu-154M EY-85691 (c/n 90A864) – likewise leased from Tajikistan Airlines – was one of four operated at various times. Here it wears the carrier's late full livery with a white fuselage. *Yefim Gordon archive*

Here EY-85691 is seen during an earlier lease, displaying an early colour scheme with a green cheatline and a white logo on a green tail. *via Chris Knott*

Egypt

Exemplifying the Egyptian flag carrier's original livery in the national colours of red, white and black, An-24B SU-AOC (c/n 57302103) was delivered in Misrair colours in 1965, by which time the carrier was already known as United Arab Airlines. This aircraft, the seventh and final An-24 delivered to Egypt, was written off in an accident at Cairo on 14th March 1970. *Yefim Gordon archive*

Egypt Air

Established on 7th June 1932 as the Misr Airwork private company ('Misr' means Egypt in Arabic), the airline came under government ownership and changed its name to **Misrair** in 1949. When Egypt and Syria formed the United Arab Republic in 1958, Misrair and Syrian Airways merged in January 1961 to form **United Arab Airlines**. The UAR (and hence the two airlines) split up in September 1961, but the Egyptian carrier retained the UAA name until October 1971 when it was renamed **Egypt Air [MS/MSR]**. The airline operated Soviet types until 1974 while President Gamal Abdul Nasser was on good terms with the USSR, but switched to Western hardware when Soviet-Egyptian relations cooled.

IL-18D SU-AOY (c/n 188011101) still wears the original United Arab Airlines green/white livery, but Egyptian Airlines titles have been added to the tail. This aircraft was returned to the USSR, serving with Aeroflot as CCCP-75429 until retirement in 1984. *Yefim Gordon archive*

IL-62 *sans suffixe* SU-AVL (c/n 00804) was one of six leased by United Arab Airlines/Egyptair in this basic pre-1973 Aeroflot livery in 1971-73. It was initially operated as CCCP-86654, regaining this identity when the lease ended. *Yefim Gordon archive*

The last-but-one Tu-154A delivered to Egyptair, SU-AXH *Ptah-Howait* (c/n 73A054), shows off its smart livery at Prague-Ruzyne. After resale to the USSR it became CCCP-85054 with the Soviet Air Force. *Martin Novak*

Cairo Charter & Cargo

Cairo Charter & Cargo was set up in 1992, taking delivery of two brand-new IL-76TD freighters and two equally new Tu-154Ms, including SU-OAC (c/n 91A898). Yet the carrier proved shortlived, selling its entire fleet to Mahan Air of Iran in March 1993; this aircraft became EP-JAZ in the process. *via Chris Knott*

Alim-Air Lines

The Cairo-based charter carrier Alim-Air Lines founded in 1996 operated three IL-62Ms leased from Uzbekistan Airways. Soon to be reregistered SU-ZDB, IL-62M UK 86932 *Mrs. Aida* (ex-Luftwaffe 11+21, c/n 3242432) is seen at Tashkent after repainting. Alim-Air Lines folded in 1997 and the aircraft were placed in storage. *Dmitriy Petrochenko*

Air Cairo (Cairo Aviation)

The charter airline **Air Cairo** founded in 1997 became the first foreign operator of the Tu-204 twinjet, taking delivery of four passenger-configured Tu-204-120s and two Tu-204-120C freighters via Sirocco Leasing. The first aircraft entered service in November 1998. In 2004 the airline was renamed **Cairo Aviation [–/CCE]**. At present most of its Tu-204s are leased out to other carriers, namely Mahan Air and TNT Aviation.

Air Cairo's first Tu-204-120C, SU-EAG (ex-Kras Air RA-64028, c/n 1450743764028), together with Tu-204-120 SU-EAF (ex-RA-64027, c/n 1450743764027) at Ul'yanovsk before delivery. *Dmitriy Petrochenko*

Air Cairo Tu-204-120 SU-EAI (ex-RA-64025, c/n 1450743164025) tucks ups up the landing gear as it completes a tight turn during a demonstration flight at the MAKS-2001 airshow at Zhukovskiy. *Yefim Gordon*

Equatorial Guinea

Ecuato Guineana de Aviacion

EGA – Ecuato Guineana de Aviación [–/ECV] was established in 1986 as the national carrier of Equatorial Guinea, operating passenger and cargo services from the country's capital, Malabo, to domestic destinations and to other countries in West Africa. The fleet consists largely of An-24 turboprops leased in Belorussia and the Ukraine as required; a Yak-40 was also reportedly operated.

An-24RV UR-26175 (c/n 77310810A) owned by the Aviant plant sits parked at the plant's Kiev-Svyatochino airfield in September 2002, still in full Ecuato Guineana de Aviacion colours. *Sergey Komissarov*

GATS Guinea

GATS Guinea [–/GTS] was one of the 'incarnations' of a carrier formed in the UAE as Gulf Aviation Technology & Services (GATS). Its fleet consisted of three IL-76TDs which had previously been on the Nicaraguan register (operating for GATS, SA) and, before that, on the Russian register. Here one of the freighters, 3C-KKF (ex-YN-CEV, ex-RA-76411, c/n 1023411384, f/n 8506) is seen undergoing maintenance at Moscow-Domodedovo on 25th November 1998. In December 2003 the airline changed its 'place of residence' once again, moving to Krghyzstan, and YN-CEV was accordingly reregistered EX-411. *Dmitriy Komissarov*

UTAGE

Union de Transporte Aereo de Guinea Ecuatorial (UTAGE) [–/UTG] based at Malabo operates Antonov turboprop airliners leased in the CIS as required. The current fleet comprises two An-24RVs leased from Moldova. This smartly painted An-26 (note the tiger tail motif) leased from Pecotox Air – apparently ER-AWV or ER-AZA – was also operated at one time. *Yefim Gordon archive*

Estonia

Estonian Air

The national flag carrier **Estonian Air [OV/ELL]** was established shortly after Estonian independence, starting operations on 1st December 1991. It was founded by the Estonian government, using the assets of the former Estonian CAD/Tallinn UAD/141st Flight – 12 Tu-134A medium-haul twinjets and four Yak-40 short-haul trijets. However, operating the Soviet jets proved difficult for economic reasons (high operating costs and expensive spares) as well as political ones. In 1995 Estonian Air leased two Boeing 737-500s to replace its ageing hardware, subsequently acquiring more 737s and a couple of Fokker 50s. By mid-1997 all four Yak-40s and ten of the Tupolevs had been auctioned off.

Tu-134A ES-AAI (c/n [83]60350, f/n 4510) in Estonian Air's simple 1990s livery is seen on approach to Moscow-Sheremet'yevo; it was sold to the Pulkovo Air Enterprise as RA-65112 in 1996. *Yuriy Kirsanov*

ELK Estonian

ELK Eesti Lennukompanii (Estonian Airways), aka **ELK Airways [S8/ELK]**, was established in 1991, operating three Tu-154Ms of the former Estonian CAD/ Tallinn UAD and an ex-Lithuanian Tu-134A from Tallinn-Ylemiste. The aircraft wore an attractive livery in the national flag colours of white, black and blue. On 31st December 2001 the airline suspended operations and the fleet was sold off.

An-26B YL-LDC (c/n 12010) was leased from Latavio in 1998, operating in this non-standard version of the ELK livery. Later the same aircraft was operated as UR-BXF until 1999. *Sergey Popsuyevich*

One of ELK Airways' three Tu-154Ms, ES-LTP (c/n 92A909), approaches Moscow-Sheremet'yevo, showing the old 'Estonian' titles. This aircraft was sold to UTair in April 2001 as RA-85727. *Sergey Sergeyev*

Tu-134A ES-LTA (c/n [83]60195, f/n 4310) taxies in at Moscow/Sheremet'yevo-1 after a flight from Tallinn. Note the different titles and the 'Flights to Moscow - Riga - St Petersburg - Turku' subtitles. *Dmitriy Petrochenko*

Enimex

Tallinn-based **Enimex [–/ENI]**, primarily a cargo carrier founded in 1994, operates an all-Soviet fleet. This included six An-28s, seven An-72s, a single An-74 and a single IL-76TD. The latter, and many of the An-72s, have now been sold off.

An-72 ES-NOB (ex-CCCP-72931, c/n 365.720.70.695, f/n 0707) was operated jointly with the British cargo carrier Channel Express for a while (note the small logo near the entry door). Enimex An-72s were often used for United Nations missions. *Yefim Gordon archive*

An-28 ES-NOD (ex-UR-28808, c/n 1AJ 002-06) was bought from Air Chaika in 1999. Later it saw service with Air Livonia. *Yefim Gordon archive*

Gambia

Gambia New Millennium Air

Appropriately registered C5-GNM, this IL-62M (c/n 3036142) was operated by Gambia New Millennium Air in 2000. Previously it had seen service with Centrafrican Airlines as TL-ACL, with Air Pass as 3D-RTI and with the State Civil Aviation Research Institute as RA-86511. *Yefim Gordon archive*

Georgia

Air Georgia

Based at Tbilisi-Lochini, Air Georgia [DA/GEO] operated three Tu-154s. One of them, Tu-154B-2 4L-85547 (c/n 82A547), is shown here departing Moscow-Vnukovo in 1998; the following year it became 4L-AAG. In contrast, Tu-154B-2 4L-85558 (later reregistered 4L-AAH) had a white livery with no cheatline and a blue logo. Both aircraft were reportedly taken over by Air Zena Georgian Airways in 1999. *Yuriy Kirsanov*

Adjarian Airlines

Adjarian Airlines [–/ADJ] of Batumi operated a pair of Tu-134AKs leased from Air Ukraine in 1997. One of them, UR 65877 (c/n [53]31250, f/n 3003), is shown here landing on runway 20 at Moscow-Vnukovo past the unfinished maintenance hangar. The colour scheme is based on Air Ukraine's 1993-standard livery. *Yuriy Kirsanov*

Batumi Adjarian Airlines

Batumi Adjarian Airlines was a different outfit operating three Tu-134B-3s bought from Lat-Charter in April-May 1997. One of them, 4L-AAD (ex-YL-LBF, c/n [03]63295, f/n 5802) is shown here landing at Moscow-Vnukovo; the aircraft still retains the ex-owner's livery with new titles/logo (it was repainted later). By 2001 all three aircraft had been sold to Russia; this one is now RA-65579 with Daghestan Airlines. *Yuriy Kirsanov*

Georgian National Airlines

Founded in 1999, Tbilisi-based Georgian National Airlines [QB/GFG] operates one IL-76TD freighter and the one-of-a-kind Tu-134B-1-3, 4L-GNA (ex-UR-BYY, ex-RA-65720, c/n [03]62820, f/n 5508). *Sergey Sergeyev*

ORBI Georgian Airways

ORBI Georgian Airways [NQ/DVU] were established in 1991 as the republic's first flag carrier, taking over most of the former Georgian CAD's assets based at Tbilisi-Lochini and Sukhumi-Babushara. Thus the original fleet comprised 12 Yak-40s of the former Tbilisi UAD/347th Flight, 15 Tu-154B-2s of the co-located 112th Flight and 14 Tu-134As divided between the 347th Flight and the Sukhumi UAD/297th Flight One of the Tu-154s was the presidential jet of Georgia's first president Zviad Gamsakhurdia, later passing to his successor Eduard Shevardnadze.

However, by October 1997 all of the original aircraft had been sold to other Georgian carriers or retired. By mid-1998, ORBI's fleet consisted of a single ex-Czech Tu-134A-1. This could not save the day and in 1999 the airline finally ceased operations.

Tu-154B-2 4L-85430 as it appeared in the days when it saw service with ORBI; the tail logo is a stylised eagle. *Yuriy Kirsanov*

The stylish ORBI livery was also worn by Yak-40s, including 4L-87305 (c/n 9320129), and several of the carrier's Tu-134As. This aircraft was taken over by Georgian Airlines after the demise of ORBI, later becoming 4L-MGC and 4L-VIP. *Yefim Gordon archive*

Tbilaviamsheni

The Tbilisi Aircraft Production Association, called Tbilaviamsheni in Georgian, operated An-26B 4L-26087 (c/n 12601) in basic Aeroflot livery with the old Georgian flag (originally it sported the non-standard registration GR-26087). The plant leased this aircraft to Air Zena Georgian Airlines in December 2003. Tbilaviamsheni also operates Tu-134AK 4L-AAJ – a former Tu-134 Balkany airborne command post transferred from the Georgian Air Force in 2003 (ex-4L-65993). *Yuriy Kirsanov*

Georgian Airlines

Tu-134A-3 4L-65798 (c/n [03]63179, f/n 5701) in Georgian Airlines' standard livery. By September 1999 it had been refitted as a government VIP aircraft for Air Zena, becoming 4L-AAI in 2000; eventually it was sold to Russia, becoming RA-65798 with East Line. In contrast, sister ship 4L-65750 had a white tail with the same logo. *via Martin Štepanek*

Part of the Georgian Airlines fleet was reregistered under the current alphabetical system. One of the aircraft in question was Yak-40 4L-87466 (c/n 9430537), which became 4L-AVD by November 2003 as shown here. However, the aircraft was no longer on the airline's fleet list at the time; some sources say it was operated by Aerovista. *via Bob Archer*

Tu-154B-2 4L-85430 (c/n 80A430) taking off at Moscow-Vnukovo illustrates the more common blue-tailed livery of the original (pre-merger) Georgian Air Lines. The black-painted dielectric panel of the Doppler speed/drift sensor just aft of the radome is unusual. *Yuriy Kirsanov*

Yak-42D 4L-TGG (ex-4L-AAM of Adjaria, c/n 4520423116579, f/n 0613) shows off the Georgian flag carrier's striking red/white livery complete with the new-style (post-'Rose Revolution') flag. *Martin Novak*

Hungary

Malev

Flag carrier **Malév Hungarian Airlines [MA/MAH]** (Magyar Légiközlekedesi Vallalat – Hungarian Air Transport Co) started life in 1954 as the successor of Maszovlet (Hungarian-Soviet Air Transport Society) founded in 1946 when Hungary fell under Soviet influence. The first Soviet hardware was the Li-2; 23 examples were operated, most of these being Li-2T transports transferred from the Hungarian Air Force after the failed anti-Communist uprising of 1956. The Li-2 served on until 1963 alongside the IL-14 (seven IL-14Ps and three IL-14Ms saw service between 1957 and 1969).

In May 1960 Malév received its first turbine-powered aircraft, introducing the IL-18 turboprop. Seven IL-18Vs and one IL-18D were delivered, the last one being retired as late as 1989; at the closing stage of the type's career it was used as a freighter. In December 1968 the carrier began replacing the IL-14 with the Tu-134 short-haul twinjet; seven 'short' Tu-134s *sans suffixe* and six stretched Tu-134As and Tu-134AKs were operated. The Tu-154 trijet replaced the IL-18 on Malév's important medium-haul routes to foreign destinations; 18 Tu-154s and Tu-154A/B/B-2s were operated over the years, including leased aircraft. A single IL-62M was briefly leased from Czechoslovakia in 1991 for inclusive tour flights to Japan. Seven Yak-40s were leased in the early 1990s for use on domestic routes.

Yet, as early as 1988 Malév began introducing Western airliners by leasing a Boeing 737-200. Like many air carriers of the former Soviet satellite states, Malév was eager to shed the socialist touch and would not operate Russian-built aircraft any more as a matter of principle. The last Tu-154s were phased out in March 2001.

A picture from the golden age of airline transport: IL-18V HA-MOE (c/n 182005505) captured by the camera shortly after landing (note that the outer engines are shut down) illustrates the original colour scheme worn by the type in Malev service. *Yefim Gordon archive*

Malev's hard-working IL-18s were relegated to the cargo role when the Tu-154 came on the scene. The same IL-18V HA-MOE is seen here in its second (1968-standard) colour scheme, with the addition of 'Air Cargo' titles. It was retired in 1987 and is now on display at the Air Force Museum in Szolnok. *Martin Novak*

Pushback time for Tu-134 *sans suffixe* HA-LBH (c/n 0350925), the last 'short' example operated by the airline (formerly Tu-134K HA-925 with the government flight). HA-LBH retained the 1968-standard livery (introduced on the Tu-134, incidentally) until it was retired to become an exhibit of the Auto & Technik Museum at Sinsheim. Note the North Korean (Chosonminhang) IL-62M in the background. *Yefim Gordon archive*

One of Malev's two initial 'glass-nosed' Tu-134As, HA-LBK (c/n 1351302); the carrier's other Tu-134As and Tu-134AKs had the 'solid' radar nose. HA-LBK stayed in service long enough to gain the current livery introduced in the 1980s but was withdrawn in 1994 after a flight incident and used for spares. *Yefim Gordon archive*

Seen here taxying out for take-off at Moscow/Sheremet'yevo-2, Tu-154B-2 HA-LCO (c/n 81A473) illustrates the carrier's current livery with additional 'Alitalia partner' titles. It now serves with Turan Air as 4K-473. *Dmitriy Petrochenko*

Atlant Hungary

Budapest-based Atlant Hungary Airlines [–/ATU] leased Tu-154M RA-85823 (c/n 88A775) from Samara Airlines In December 2000. The aircraft was reregistered HA-LGA, operating in basic Samara colours between 28th March and 26th June 2001. After that the airline ceased operations and the aircraft was returned, regaining its former identity; the plans to lease three further Tu-154Ms never materialised. *via Martin Štepanek*

Linair

Linair Hungarian Regional Airlines [–/LIN] operated a pair of Yak-40s on behalf of the Hungarian Airport Administration. One, registered HA-LRA (ex-ČSA Czechoslovak Airlines OK-EED, c/n 9440837), was obtained from Aviaexpress (see next page); it was outfitted as a Yak-40 'Salon 1st Class' VIP jet and used jointly with Malev from August 2000 until retired in June 2001. The other, HA-YLR, was a navaids calibration aircraft; it was likewise retired in 2001 and is preserved at Budapest-Ferihegy. *via Bob Archer*

Pannon Airlines

Another short-lived Hungarian Tu-154M operator was Pannon Airlines [–/PHP] which leased Tu-154M RA-85650 (c/n 88A788) from the Vnukovo aircraft repair plant in May 1993 as HA-LCX. The jet was operated jointly with the travel agency Jorgos Travel whose titles were applied later. The carrier intended to lease a second Tu-154M but went bankrupt before it could do so; in 2002 HA-LCX was returned to the lessor, subsequently becoming LZ-LCI. *via Martin Štepanek*

Aviaexpress

Aviaexpress [RX/AEH] operated Yak-40s HA-LRA and HA-YLR on behalf of the Hungarian Airport Administration before turning them over to Linair Hungarian Regional Airlines (see previous page). Additionally, Aviaexpress leased Yak-42D UR-42544 from Air Ukraine in May 1993. *Martin Novak*

Cityline Hungary

Cityline Hungary Kft [ZM/CNB] was established in 2003 as a cargo carrier operating An-26s from Budapest-Ferihegy. Four aircraft are currently on strength; three of them, including HA-TCP (ex-Kroonk Airlines UR-26129, c/n 12703), are ex-Ukrainian aircraft, while the fourth was bought in Moldova. *Martin Novak*

Indonesia

Lion Airlines

Lion Airlines [JT/LNI] based at Jakarta operate scheduled and charter services linking Indonesia with various parts of the world, using Boeing 737-300s/-400s, McDonnell Douglas MD-82s and MD-90s. However, the airline wet-leased three Russian-registered Yak-42Ds from the Chelyabinsk Air Enterprise in 2001-02, operating them in partial Lion Airlines colours on the Jakarta-Kuala Lumpur service.

An atmospheric shot of Yak-42D RA-42412 (c/n 4520422219055, f/n 0415) staying overnight at Moscow-Domodedovo in 2002. The aircraft still wears the livery of Lion Airlines and the name *Batavia* but is already operated by the real owner, the Chelyabinsk Air Enterprise. Sister ship RA-42387 leased from the same carrier was named *Pontianak*.
Dmitriy Petrochenko

Iran

Iran Air Tour

Iran Air Tour [B9/IRB] was established in1992 as a subsidiary of flag carrier Iran Air operating exclusively Soviet/Russian airliners on scheduled domestic services and international charters. The airline's main bases are at Tehran and Mashhad, with a secondary hub at Mehrabad. The Tu-154M is the mainstay of the fleet, a steady succession of these aircraft being leased from Russian airlines as required; 12 were on strength in August 2006. Several Yak-42Ds were also leased in the early 1990s but apparently found to be too small. Until 2006, Iran Air Tour had no livery of its own, operating its aircraft in the basic colours of the lessors.

In 2006, after seven years of negotiations, Iran Air Tour placed an order for five new Tu-204-100s due for delivery before the end of 2008. Thus it will be the first foreign operator of the baseline version powered by Russian PS-90A engines.

Until 2006, Iran Air Tour had no livery of its own, operating its Tu-154Ms in the basic colours of their proper owners. EP-MBJ (ex/to RA-85801, c/n 93A960) wore the basic colours and even the logo of Kogalymavia; it was leased from May 2000 to May 2002. *Yefim Gordon archive*

By comparison, EP-MBB (ex/to RA-85830, c/n 89A821), operated from October 1998 to October 2002, retained the basic livery of Omskavia; this is known colloquially as *Zmey Gorynych* (= dragon in Russian folklore)! *via Viktor Kravchenko*

Kish Air

Another major Tu-154 operator in Iran is the charter carrier Kish Air [Y9/IRK] established in 1989 as a subsidiary of the Kish Development Organisation (a free zone enterprise on Kish Island). Again, it leases the jets from Russian air carriers as required. After a period of operating the airliners in the lessors' basic liveries Kish Air introduced several colour schemes of its own. EP-LBN bought from Sayakhat in 1999 (ex-UN 85836, c/n 85A717) illustrates one of the versions; it was later sold to Russia as RA-85081 No 2. EP-LBR was painted similarly but the Kish Air titles were in a different font. *via Viktor Kravchenko*

Mahan Air

Mahan Air [–/IRM] based in Kerman bought up the entire fleet of the defunct Cairo Charter & Cargo in March 1993, including the two IL-76TDs. Here, EP-JAY (ex-SU-OAA, c/n 1013409297, f/n 8305) approaches Moscow-Sheremet'yevo; the livery is little changed from that of the ex-owner. Note the registration is repeated on the outer engines. Both freighters went to Qeshm Air in 2000, this aircraft being reregistered EP-TQJ. *via Viktor Kravchenko*

Seen here landing at Moscow-Vnukovo after a post-overhaul checkout flight, Tu-154M EP-ARG is also an ex-Egyptian aircraft (ex-SU-OAD, c/n 92A899). This bird was sold to Caspian Airlines in 2000 as EP-CPO. However, other Tu-154Ms were operated later. *Dmitriy Petrochenk*

Payam Air

Based at Karaj-payam International airport, cargo carrier Payam Air [–/IRP], formerly IPTAS (Iran Postal & Telecommunications Aviation Services), leased IL-76s from CIS airlines as required. At least ten aircraft have been operated over the years. Here, IL-76TD EP-TPV (ex-RA-76475, c/n 0043451523, f/n 3901) shows off the basic livery and tail logo of Samara Airlines from which it was leased in 2000. *via Viktor Kravchenko*

Qeshm Air

Qeshm Air [–/IRQ] operates passenger service from Qeshm Island (a Free Zone island at the mouth of Persian Gulf) to cities in mainland Iran and also the United Arab Emirates, as well as cargo service to other countries. Its Qeshm-Dubai service is popular with foreign guest workers living in the United Arab Emirates. Currently the fleet consists of two IL-76TD freighters.

Yak-40 EP-TQH (ex-EY-87217, c/n 9510340) was leased from Tajikistan Airlines in 1999-2000, hence the blue/white livery. *Sergey Popsuyevich*

IL-76TD EP-TQI (ex-Mahan Air EP-MAH, ex-SU-OAB, c/n 1013409321, f/n 8401) is about to depart from Moscow/Sheremet'yevo-2. *Yuriy Kirsanov*

Kampuchea (Cambodia)

Imtrec Aviation

Founded in 2000, Imtrec Aviation [–/IMT] based at Phnom Penh operates three An-12 freighters and this Laotian-registered IL-76TD 'Falsie', RDPL 34146 (possibly ex-IL-76MD RA-76578, c/n 0043449468, f/n 3707). Here the aircraft is pictured undergoing maintenance at Sharjah; the tail gunner's station is just discernible. At least one more example, RDPL-34148, is also in service with the carrier. *Sergey Popsuyevich*

One of Imtrec Aviation's three Kampuchean-registered An-12s, civil-configured An-12BP XU-365 (ex-Avial' RA-11899, c/n 402601). The aircraft also carries the titles of the Bangladeshi carrier Bismillah Airlines with whom it is jointly operated. This is the second An-12 to have this registration; the original XU-365 (c/n 01348005), a 'military' An-12BP also operated by Imtrec Aviation, was sold as EX-034 (see page 114). *Yefim Gordon archive*

Kampuchea Airlines [E2/KMP] received two Tu-134AK VIP jets in the mid-1980s. XU-101 (ex-CCCP-65065 of the Soviet government flight, c/n [73]49890, f/n 4007) was a 'glass-nosed' example; XU-102 had a 'radar nose'. Both were sold to Nizhegorodskie Airlines in 1995 as RA-65065 and RA-64451. *Yefim Gordon archive*

Yana Airlines

Yana Airlines acquired two IL-62Ms from the Liberian carrier Air Cess in 1998. One of them, XU-229 (ex-OK-OBL of Egretta-BMI, c/n 4445032) is shown here at Sharjah where it was based. The latter fact suggests that the Kampuchean registration was just a 'flag of convenience'. *Martin Novak*

Kazakhstan

Kazakhstan Airlines

Kazakhstan Airlines/Kazakhstan Aue Zholy or **Kazair [K4/KZA]** was this Central Asian republic's first flag carrier formed in 1991 as the successor of Aeroflot's Kazakh CAD. The airline was headquartered in Almaty, which was then the republic's capital. The large fleet distributed between several divisions included at least one An-12, at least 36 An-24s based in Akmola, Burundai, Goor'yev, Karaganda, Kustanay and Shimkent, at least ten An-26s, nine An-30s based at Burundai, four IL-76TDs, seven IL-86s, nine Tu-134As at Almaty (all based at Almaty), 26 Tu-154B-2s and Tu-154Ms based at Almaty and Karaganda, at least 30 Yak-40s based in Aktyubinsk, Zhambyl, Kzyl-Orda, Pavlodar, Taldy-Kurgan and Ust'-Kamenogorsk, and at least six Yak-42Ds, plus Mi-2 and Mi-8 helicopters based at Burundai, Goor'yev and Semipalatinsk. Kazakhstan Airlines operated a wide network of scheduled and charter services to destinations within the republic, other CIS states and Western Europe. In order to add passenger appeal and facilitate long-haul services the airline intended to order four Boeing 767s. However, Kazakhstan Airlines' financial position was rather shaky; during the five years of its existence the airline incurred a debt of 19 billion tenge (more than US$ 180 million). Finally, on 20th August 1996 Kazakhstan Airlines filed for bankruptcy. Rather than attempt to rescue the ailing giant, the government chose to liquidate the airline; this led to the emergence of a new flag carrier, Air Kazakstan (see page 62), and several other new airlines which had previously been Kazair's territorial divisions.

Seen here on approach to Moscow-Domodedovo in the mid-1990s, Tu-134A-3 UN 65147 (c/n [93]61012, f/n 5002) wears the full colours of Kazakhstan's original flag carrier with Kazair and Kazakhstan Aue Zholy (Kazakhstan Airlines) titles. *Yuriy Kirsanov*

Another visitor to Domodedovo. Tu-154B-2 UN 85537 (c/n 82A537) wears an unusual livery with only a pinstripe in lieu of a cheatline and English titles. *Yuriy Kirsanov*

Seen here during a visit abroad, IL-86 UN-86116 (c/n 51483209084) still wears Aeroflot titles, the registration and the Kazakh flag on the tail being the only evidence of its new ownership. Note that the Soviet CCCP- prefix is still carried on the wings. *Yefim Gordon archive*

Caught by the camera on rotation on Domodedovo's runway 14R, IL-86 UN 86071 (c/n 51483204038) is a good example of how Kazair's aircraft usually looked in the early 1990s – with full tail colours and titles but an Aeroflot cheatline. *Yuriy Kirsanov*

Seen on short finals to Moscow-Domodedovo, An-24B UN-46342 (c/n 97305610) is still in basic Aeroflot colours. Note the unusually small size of the UN- prefix. *Yefim Gordon archive*

Basking in the sun on the east apron at Domodedovo, Yak-42D UN42446 (c/n 4520423308017) was delivered new in this somewhat non-standard full Kazair livery (note the different tail colours and logo). It became RA42446 by 2000 and was operated as a VIP jet by Vnukovo Airlines for the Russian Federal Government Communications and Information Agency (FAPSI). *Yefim Gordon archive*

UN 76435 (c/n 1023413428, f/n 8607) basking in the afternoon sun is the only confirmed example of Kazair's four IL-76TDs to receive the revised livery. Tragically, this aircraft was lost on 12th November 1996 when it collided with Saudi Arabian Airlines Boeing 747-186 HZ-AIH near New Delhi, killing 349 – the worst-ever mid-air collision in commercial aviation history. *via Viktor Kravchenko*

Air Kazakstan

The new flag carrier that picked up the ball from the defunct Kazakhstan Airlines was called **Air Kazakstan [9Y/KZK]**. The carrier inherited most of Kazair's fleet, including all the big jets, starting operations on 10th March 1997. Yet the story was repeated: in February 2004 Air Kazakstan ceased operations and was declared bankrupt in April, ceding the flag carrier role to Air Astana based at the new capital, Astana. Here, Tu-154B-2 UN 85240 (c/n 77A240) vacates runway 14L at Moscow-Domodedovo upon arrival in 2001, showing Air Kazakstan's full livery with no cheatline. This aircraft was retired by 2002. *Dmitriy Petrochenko*

Atyrau Airways

Established in 1996, **Atyrau Airways**, aka **Atyrau Aue Zholy [IP/JOL]**, based in Atyrau (formerly Goor'yev) took over the An-24s and Mi-8 helicopters of the local Kazair branch. The turboprops have now been withdrawn, but in 1998 the airline started jet operations, consecutively buying four Tu-134A/AKs from the Russian carrier Komiinteravia. One Tu-154B-2 and two Tu-154Ms have also been operated; of these, only one Tu-154M remains in service.

Tu-134A-3 UN-65070 *Tungysh* (ex-Komiavia RA-65070, c/n [73]49912, f/n 4104) looks immaculate in Atyrau Airways' current colours with a 'waves' motif on the nose, no cheatline, English-only titles and a very bold tail logo. *aviaphoto.ru*

Tu-154M UN 85855 (ex-China Xinjiang Airlines B-2621, c/n 89A823) seen here at Astana exemplifies the previous livery; unusually, the titles are in the Kazakh language only (Atyrau Aue Zholy). Purchased from Sayakhat in 2003, this aircraft has now been sold to Mak Air. *Aidan Curley*

SAN

SAN Air Company [S3/SND], a Kazakh-German joint venture based in Karaganda, operated two Tu-134As leased from Tajikistan Airlines in 1997-99 and four Tu-154Ms leased from Bashkirian Airlines and Samara Airlines in 1996-99. Tu-134A-3 EY-65788 (c/n [03]62835, f/n 5509) is shown here; note the 'German flag' flash near the wing root indicating the airline's German affiliations. *Dmitriy Petrochenko*

SCAT

SCAT Joint-Stock Co [DV/VSV] established in 1997 is a major regional airline based in Shymkent. The fleet includes 28 An-2 utility biplanes and 23 An-24 airliners (15 An-24Bs and eight An-24RVs); the turboprops are used on both domestic and international services. One of the An-24Bs, UN 46438 (c/n 87304309), is seen here visiting Astana. Additionally, SCAT briefly operated a single Yak-42D trijet. *Aidan Curley*

Sayakhat

Established in 1989, **Sayakhat [Q9/SAH, later W7/SAH]** (the name means 'journey') started operations in 1991 as the first private airline company in Kazakhstan. For the first seven years it was a cargo carrier operating five IL-76TDs from Almaty. For a while they were jointly operated with the Russian airline Tesis. In 2002 the carrier diversified, launching passenger services with seven ex-Chinese Tu-154Ms. Four IL-76s and three Tu-154s are still in service.

Sayakhat's IL-76TDs are regular visitors to Moscow/Vnukovo-1. UN-76384 (c/n 1033401015, f/n 7604) exemplifies the original livery, which is rather more appealing than the nondescript current grey/white colours. *via Viktor Kravchenko*

Kyrghyzstan

Kyrghyzstan Airlines

Kyrghyzstan Airlines/Kyrghyzstan Aba Zholdoru [K2/KGA, later R8/KGA], the republic's flag carrier, was formed from Aeroflot's Kirghiz CAD in 1992. Headquartered at Bishkek (formerly Frunze), the Kyrghyz capital, it took over the solitary IL-76TD, six Tu-134As, five Tu-154B-1s, seven Tu-154B-2s and two Tu-154Ms of the Frunze UAD, the 11 Yak-40s of the Frunze and Osh UADs, as well as at least one An-26 freighter, a number of An-28 turboprop feederliners, Mi-2 and Mi-8 helicopters.

The carrier operates scheduled domestic and international services, the latter including New Delhi, Dushanbe, Jalalabad, Moscow, Novosibirsk, Osh, Seoul, Sharjah, Urumqi and Yekaterinburg.

The current fleet includes three Tu-134A-3s, three Tu-154B-1s, four Tu-154B-2s, two Tu-154Ms, 13 Yak-40s, six Mi-2s and five Mi-8Ts; the single leased Airbus A320 has been returned.

The airline also performs government VIP flights. The government holds an 81% stake in the assets.

The gear retraction sequence almost complete, Tu-154M EX-85718 (c/n 91A900) in the old colours of Kyrghyzstan Airlines makes a smoky departure from Moscow-Domodedovo. The titles on the port side read 'Kyrghyzstan Aba Zholdoru'. *via Viktor Kravchenko*

Seen here during a visit to Salzburg, Tu-154B-2 EX-85519 (c/n 81A519) used as a government VIP jet exchanged the original standard Aeroflot-style livery for an overall white finish with red KYRGHYZSTAN titles and tail logo. *via Martin Štepanek*

Tu-154M EX-85762 (c/n 92A945) seen here taxying at Moscow-Domodedovo in August 2002 shows off the current Kyrghyzstan Airlines livery introduced in late 1998 with the A320. *Yuriy Kotel'nikov*

Tu-134A EX-65779 (c/n [93]62602, f/n 5406) taxying for take-off from Moscow-Domodedovo's runway 32R shows that the titles on the starboard side are in English; compare the spelling to the IL-76 at the foot of this page, though. *Yuriy Kirsanov*

The fleet of Kyrghyzstan Airlines included this single IL-76TD transport, EX-76815 (c/n 1013409310, f/n 8308). By 2002 the aircraft had been placed in storage; eventually it was sold and is now operated by Imtrec Aviation of Cambodia with the Laotian registration RDPL-34148. *Chris Doggett*

Altyn Air

In 2002-05 Tu-154M EX-85718 was leased to Bishkek-based Altyn Air [QH/LYN], operating in the airline's striking red/white livery; *altyn* means 'gold' in the Turkic languages. Apart from scheduled and charter services, EX-85718 occasionally performed government VIP flights. *Eero Saarela*

Altyn Air's fleet also includes this Tu-134A with the current-style three-digit registration EX-020 (ex-Georgian Airways 4L-65750, c/n [93]61042, f/n 5005) and a pair of An-24RVs. *J de Raaf*

Aerovista

Although listed as a Kyrghyz airline, Aerovista [–/AAP] is based in Sharjah, UAE. The constantly changing fleet currently includes three An-24s, one An-26s, one IL-18V and seven Yak-40s. An-24RV EX-24807 (ex-S9-CDA, TAROM YR-BMM, c/n 77310807) is no longer in use, having been sold to Serendib Express. Aerovista has suffered the ignominy of being placed on the JAA's list of airlines banned from flying to Europe due to unsatisfactory operating standards. *Yefim Gordon archive*

Despite its Sudanese registration, this convertible An-26B-100 ST-JAC (ex-An-26 EX-003, ex-RA-26019, c/n 07310203) was also operated by Aerovista. Note the logo of the Irkutsk Aircraft Repair Plant No 403 (which performed the conversion) on the nose, the integral airstairs (a feature of the An-26B-100 version) and the display placards near the entry door showing possible cabin configurations.
Yefim Gordon archive

Phoenix

Based at Sharjah, UAE, **Phoenix Aviation [P3/PHG]** is a rather obscure airline. Originally listed as a UAE carrier, it is now reported for Kyrghyzstan and has a second base at Bishkek-Manas. In recent years it operated at least one An-12, ten IL-18s and three IL-76TDs registered in various states. Currently, however, Phoenix Aviation has an all-737 fleet.

Wearing Phoenix Airlines titles over what was originally the livery of the Russian airline Ramair, IL-18D EX-75466 (c/n 187010403) is seen during one of its frequent visits to Sharjah – still in standard configuration. It was converted to an IL-18GrM in 2001, being refitted with a cargo door sourced from the crashed RA-75554. It now serves with Anikay Air (see page 69). *Yefim Gordon archive*

In the late 1990s Phoenix Aviation was headquartered in the UAE. This night scene at Sharjah in 1999 shows IL-76TD ST-AQA (ex-RA-76498, c/n 0023442218, f/n 3105) in the basic livery of ex-owner Moscow Airways with the titles of Phoenix Aviation and its Sudanese partner Trans ATTICO. *Peter Davison*

Another Phoenix Aviation IL-76TD placed on the Bosnian register as T9-CAB (ex-T9-QAB, ex-Continental Airways RA-76355, c/n 1023408265, f/n 8207) in basic Ramair colours at the same location. *Peter Davison*

One more Phoenix IL-76TD, ST-AQB (ex-Atlas Air EP-ALA, c/n 0053460795, f/n 4509) in the grey/white colours of Atlas Air and again with additional Trans ATTICO titles. *Yefim Gordon archive*

British Gulf International

British Gulf International Airlines [–/BGK] is another 'expat airline'. Although based at Sharjah, it is listed as a Kyrghyz carrier and the entire fleet consisting of six An-12s is on the Kyrghyz register.

Demilitarised An-12BP EX-164 *Alex* (ex-RA-11002, c/n 5343703) retains basic Aeroflot colours. *Yefim Gordon archive*

EX-161 *Fatima* (ex-S9-BOT, c/n 5343305), another demilitarised An-12BP, has a more fanciful livery with a blue tail and no cheatline. As the AI-20 turboprop is a smoky engine, the An-12's rear fuselage is typically black with soot, especially on the port side, and the soot is damn hard to get rid of, creeping into the rivet joints and highlighting the fuselage frames as seen here. *Yefim Gordon archive*

Anikay Air

Founded in 2003, Anikay Air [–/AKF] is a small cargo airline operating from Bishkek-Manas and Fujairah (UAE). Apart from a single Boeing 727-200C, the fleet comprises three IL-18s, including IL-18E EX-601 (c/n 185008601). The green/white livery betrays its previous ownership by Santa Cruz Imperial with whom it served as EL-ALD. *Sergey Popsuyevich*

In contrast, this rare IL-18GrM freighter with a large side cargo door, EX-75466, is in the basic colours of ex-owner Phoenix Aviation (see page 67). *Sergey Popsuyevich*

Itek Air

The fleet of Itek Air [–/IKA], based at Bishkek, consists of a mere two aircraft. One of them is Tu-134A-3 'Salon' EX-65119 (c/n [83]60475, f/n 4608) seen here taxying for take-off from runway 32R at Moscow-Domodedovo. This aircraft is now on lease to Benin Golf International. *Yuriy Kirsanov*

Itek Air's other aircraft is this Tu-154B-2, EX-85369 (c/n 79A369) – a fine sight as it departs Domodedovo on a sunny autumn day. *Dmitriy Petrochenko*

Quadrotour Aero

Quadrotour-Aero was an obscure holiday airline whose only hardware was this IL-62 *sans suffixe* with the non-standard registration EX-62100 (ex-UN-86503, ex-TAROM YR-IRD, c/n 51903). Quadrotour-Aero bought the jet from the defunct Kazakh airline Trans Asian Airlines (which had been banned from European skies) in 2000, retaining the basic TAA livery, and tried flying to Europe! Yet, murder will out, and the new operator was blacklisted as well. *via Viktor Kravchenko*

Laos

Lao Airlines

Founded in 1979 as Lao Aviation [QV/LAO], the Laotian flag carrier operated such Soviet airliners as the An-24RV and Yak-40, as well as Mi-8 helicopters. These were gradually augmented/replaced by the Xian Y7 (a Chinese An-24 copy) and Harbin Y12 II, as well as Western types.

RDPL-3.4006 (c/n 67310610) was the second of the original six Soviet-built An-24RVs delivered in 1976 and 1977; it had been retired by 1997. *Yefim Gordon archive*

Astro Air

Astro Air operated an IL-76TD 'Falsie' registered RDPL-34141 (ex-Aram Air EP-RAB, c/n 0053465941, f/n 4906) in 2002. Unfortunately this aircraft crashed on landing at Baucau, East Timor (Indonesia), on 31st January 2003. *Yefim Gordon archive*

Latvia

Latavio

The original flag carrier, **Latavio Latvian Airlines** (Latvijas Aviolinijas) **[PV/LTL]**, was established in 1991 after the republic broke away from the Soviet Union. It inherited the assets of the former Latvian CAD – 30 An-2 utility biplanes based at various locations, eight An-24Bs, eight An-24Bs, four An-24RVs and three An-26Bs of the Riga UAD/106th Flight based at Riga-Spilve, ten of the 14 Tu-134B-3s operated by the Riga UAD/280th Flight based at Riga-Skul'te, seven Tu-154s operated by the same flight, 20 Mi-2 helicopters and a single Yak-40. The airline operated domestic services and international flights to destinations in Eastern and Western Europe. It also wet-leased its aircraft to other carriers, such as Harco Air Services of Nigeria.

Since the relations between Russia and the Baltic States were rather strained in the early 1990s, Latavio soon ran into maintenance problems with the Russian hardware. Part of the fleet was sold (for instance, four of the Tu-134s were bought by Latvian charter carriers Lat-Charter and Baltic Express Line, while some of the Mi-2s were taken over by the Latvian Air Force). Some aircraft were cannibalised for spares to keep the others flying. Eventually, however, Latavio was forced to cease operations in 1996, selling the jet airliners to CIS operators; the Antonov fleet was partly sold fo the Air Force and foreign operators and partly retired. An airline called airBaltic became the new flag carrier.

Latavio Tu-134B-3 YL-LBA (ex-CCCP-65146, c/n [93]61000, f/n 5001) makes a splendid picture as it comes in to land at Moscow-Sheremet'yevo with aft trim very much in evidence. Note the black-painted wings and tailplanes, the tips of the main gear fairings painted Dayglo orange to avoid damage by ground vehicles and the differently coloured mainwheels. In 1995 the aircraft was impounded at Sheremet'yevo and eventually sold to Alrosa as RA-65146 in 1999. *Yuriy Kirsanov*

Still wearing an experimental Aeroflot livery with a blue/white tail, Latavio Tu-154B-2 YL-LAH (ex-CCCP-85558, c/n 82A558) taxies out from Moscow/Sheremet'yevo-1 past the characteristic 'mushroom' concourse. It received full Latavio colours before being sold to Georgian Airlines in 1993 as 4L-85558. *Sergey Sergeyev*

Lat-Charter

Lat-Charter's original fleet consisted of six Tu-134B-3s, nearly all of which sported different versions of the same livery. YL-LBG (ex-CCCP-65669, c/n [03]63333, f/n 5806) is illustrated here. Note the window inserts making the windows appear rectangular from the inside; the aircraft was outfitted with a new interior by Israel Aircraft Industries' Bedek division. This aircraft is now operated by GST-Aero as UN 65699. *Sergey and Dmitriy Komissarov collection*

By 2003 all of the Tu-134Bs had been sold. In 2001 Lat-Charter acquired two Yak-42Ds from China General Aviation, including YL-LBU (ex-B-2757, c/n 4520423403018). These wore a totally different livery as seen here. YL-LBU is now UR-KBR with an as yet unidentified Ukrainian airline. *via Bob Archer*

RAF-Avia

RAF-Avia [–/MTL] was established on 31st October 1990 as the flight unit of the RAF minibus factory in Jelgava near Riga (Rigas Automobiles Fabrika). Its function was to deliver parts supplied by subcontractors from all over the former USSR. The first of an initial three An-26 freighters arrived on 31st May 1991, flying the inaugural service on 19th August.

In December 1994 RAF-Avia was transformed into a joint-stock company as a result of the parent company's re-organisation. It continued to co-operate with the factory but also performed commercial transportation services. The RAF minibus factory subsequently went bankrupt.

Privatisation of the company began in 1996 and was completed in November 1999. RAF-Avia continued expanding its fleet, increasing the number of An-26s to five and replacing two of the original three with newer examples. In 2001 the airline acquired a single An-74 which was upgraded to An-74TK-100 convertible configuration, enabling passenger services. In 2005 RAF-Avia added a Western aircraft (a SAAB 340A Cargo conversion) to its fleet.

RAF-Avia is a partner of the TNT, Federal Express, United Parcel Service and DHL Worldwide Express delivery companies, performing overnight cargo flights. It also performs contract work for the Latvian Ministry of Defence/NATO and the United Nations.

In 2002 the airline entered strategic agreements with Seagle Air and Slovakian Airlines to provide cargo transport and scheduled passenger services.

RAF-Avia's small fleet includes a single An-74TK-100 registered, very appropriately, YL-RAF (ex-Aviacor RA-74025, c/n 365.470.95.905, f/n 1407); this aircraft has been converted to TK-100 configuration. *Yefim Gordon archive*

An-26 YL-RAA No 2 (ex-RA-26064, c/n 17311206) comes in to land on runway 22L at Copenhagen-Kastrup on a July afternoon in 1998, showing RAF-Avia's early colours; note the white engine nacelles and additional DHL titles. It was followed a few minutes later by sister ship YL-RAC with grey nacelles. The original YL-RAA (ex-CCCP-29107, c/n 6210) had been sold to the Congo back in 1997. *Dmitriy Komissarov*

An-26B, YL-RAE (ex-CCCP-26200, c/n 14004), illustrates RAF-Avia's latest livery. *Martin Novak*

YL-RAF

YL-RAA
DHL

YL-RAE

YL-RAA No 2 as it looks today – a one-off variation on the current livery seen on the previous page with a silver-painted tail and blue engine cowlings. *Yefim Gordon archive*

Liberia

Santa Cruz Imperial

Sharjah-based **Santa Cruz Imperial [–/SCI]** founded in 1996, originally known as Flying Dolphin, was primarily a cargo airline flying mostly Antonov types.

An-8 EL-AKZ (ex-RA-69353, c/n 0E 3420) in basic Santa Cruz Imperial colours sits in storage at Sharjah, awaiting scrapping. *Yuriy Kirsanov*

The carrier's other An-8, EL-AKY (ex-J2-KBG, ex-RA-69343, c/n 0G 3410), was in basic Aeroflot colours. *Yefim Gordon archive*

An-24B EL-ALF (ex-UN 26200, c/n 14004), illustrates RAF-Avia's latest livery. *Martin Novak*

EL-ALJ (ex-Pamir Air YA-PAB, ex-Air Cess EL-AKW, ex-RA-12191, c/n 8346202) was originally an An-12BK but had the large Initisiativa radar replaced with an ROZ-1 radar in a smaller radome. It was one of the nine An-12s operated over the years and is seen here undergoing maintenance. In addition to the Antonovs, five IL-18s were operated. *Martin Novak*

Libya

Libyan Air Cargo

At a glance, this may seem to be an ordinary IL-76M operated by Jamahiria Air Transport, as Libyan Air Cargo was known previously. Yet the refuelling pods under the wings and on the rear fuselage tell a different story: 5A-DLL No 2 (c/n 0093493799, f/n 7010) is an IL-78MKE tanker/transport! *Sergey Sergeyev*

For many years Jamahiria Air Transport's IL-76s, including IL-76TD 5A-DNS (c/n 0023439145, f/n 2907) seen here visiting Prague, were white overall with a small Libyan Arab Airlines tail logo. *Martin Novak*

Despite being registered in Equatorial Guinea, An-72 3C-QTA (probably ex-RA-72929, c/n 365.720.60.653, f/n 0610) seen at Malta-Luqa is operated for the Libyan government in basic Libyan Air Cargo colours, as indicated by the Libyan flag aft of the flightdeck. *Yefim Gordon archive*

Smartly painted in full Libyan Air Cargo colours, An-124-100 5A-DKN *Sabrata* (ex-Antonov Airlines UR-82003, c/n 19530502792, f/n 0302) undergoes maintenance in the Aviant hangar at Kiev-Svyatoshino, surrounded by An-32 airframes awaiting completion and an An-12 due for refurbishment. *Sergey Popsuyevich*

Air Libya

Founded in 2002 as Air Libya Tibesti [7Q/TLR], this airline, now known simply as Air Libya, leased Yak-40 EX-87426 (c/n 9420235) from Aerovista that year. The jet is now registered 5A-DKK, inheriting the identity of a Jamahiria Air Transport IL-76M destroyed in 1986. Tu-154B-2 4L-85496, leased from Georgian Air Lines from 2003 to 2005, wore a similar livery. *via Martin Štepanek*

In contrast, VIP-configured Yak-40KD EX87227 (*sic*, c/n 9841659) retains the basic colours of Phoenix Airlines. Note the variations in the titles ('Tibesti Air Libya' or 'Air Libya Tibesti'). The Yak-40 is the airline's primary type, at least six being operated. *Yefim Gordon archive*

Sin Said Airlines

Founded in 2000, Sin Said (an airline associated with Centrafrican Airlines) operated various Soviet types from Ras al Khaimah (UAE) and Tripoli. IL-76TD 5A-DKS (ex-Aviaenergo RA-76843 (c/n 1033418584, f/n 9006?) wears the Sin Said blue/yellow/green cheatline and titles but no tail logo. The airline also had three VIP-configured IL-62Ms. *Yefim Gordon archive*

An-72 EL-ALL (ex-Enimex ES-NOL, c/n 365.720.80.789, f/n 0809) seen at Sharjah in 1999 is the reverse case, with a Sin Said tail logo but no titles. Note that the sliding rear cargo door segment is partially open and the flightdeck glazing is covered to keep the flightdeck cool. The aircraft was later reregistered 3C-QQS and operated by West Africa Air in the same livery for a while before reverting to its Estonian registration. *Sergey Popsuyevich*

Libyan Government

Anonymous-looking IL-62M 'Salon' 5A-DKT *Rayane* (ex-3C-QQR, c/n 4648414) tucks up its landing gear as it departs from Moscow-Domodedovo. *Sergey Krivchikov*

Lithuania

Lithuanian Airlines

Lithuanian Airlines (LAL) [TE/LIL] were re-established as the newly-independent Lithuania's flag carrier in 1991, becoming the successor of the pre-war Lietuvos Oro Linijos. The carrier inherited most of the Lithuanian CAD/Vilnius UAD's fleet – two An-24Bs, two An-24RVs, three An-26Bs, nine Tu-134As and 12 Yak-42s, as well as a sizeable number of An-2s based at Klaipeda. The Yak-40 trijets, also based at Vilnius, passed to a different operator, although one of them did briefly wear the early LAL livery – applied by mistake.

The airline performed scheduled and charter flights throughout Europe. As early as 1991, however, LAL started re-equipping with Western types (Boeing 737-200s, followed by more modern 737-500s, SAAB 340s and SAAB 2000s. Hence the Soviet types were relegated to charter duties and withdrawn altogether by 2000, the jets going to the CIS and the propliners being sold to domestic operators (including Air Klaipeda and the Lithuanian Air Force).

Fresh from overhaul, An-24B LY-AAF (ex-CCCP-46320, c/n 97305406) demonstrates Lithuanian Airlines' early 1993-standard livery as it taxies as Kiev-Zhuliany. The aircraft was sold to Federal Airlines of Sudan as ST-AOE in 1998. *Popsuyevich*

A typically smoky approach by Tu-134A LY-ABF (ex-CCCP-65088, c/n [73]60172, f/n 4307) wearing LAL's 1994-standard livery. This aircraft was sold to Avialeasing Co as Tu-134A 'Salon' RA-65088 in 1996. *Yuriy Kirsanov*

In the carrier's early days Lithuanian Airlines' aircraft were operated in basic Aeroflot colours without titles and with a Lithuanian flag on the tail. A good example is Yak-42 LY-AAW (ex-CCCP-42359, c/n 4520423811417, f/n 0409); this aircraft was sold to Domodedovo Airlines as RA-42359 in 2002. *via Martin Štepanek*

Yak-42 LY-AAP (ex-CCCP-42342, c/n 4520421706302, f/n 1007) in 1994-standard colours is seen on final approach to Frankfurt-Main. This aircraft was sold to Irtysh-Avia of Kazakhstan as UN 42342 in 1998; after a series of leases to Africa with this registration it became EK-42342 with Aero Service in 2002 and is now operated by Interavia as RA-42342 since 2006. *Helmut Walther*

Lietuva (Air Lithuania)

Kaunas-based Lietuva [TT/KLA], aka Air Lithuania, took over a single Tu-134A (LY-ABA, c/n 3352003) from Lithuanian Airlines; it is seen here sharing the apron at Malta-Luqa with a sister ship belonging to Balkan Bulgarian Airlines. The aircraft was sold to Air Tatarstan in 1999 as RA-65973. *Etienne Zammit*

The Yak-40 constituted the bulk of Lietuva's fleet. Executive-configured Yak-40 'Salon 2nd Class' LY-AAD (ex-CCCP-87388, c/n 9412032) exemplifies the early livery. After a lease to White Eagle Aviation as SP-FYS in 1997-99 it was returned and then sold to Tulpar Airlines in 2002 as RA-887388. *via Chris Knott*

Here, Yak-40 'Salon 2nd Class' LY-AAA (ex-CCCP-88279, c/n 9720154) is shown in the carrier's more stylish late livery with 'Air Lithuania' titles. This aircraft was sold to the Ukraine in early 2002, becoming UR-LEV. All of the carrier's six Yak-40s had been disposed of by 2004. *via Chris Knott*

Lietuva
TU-134A
LY-ABA

Lietuva
LY-AAD
JAK-40
TORA 3050

Air Lithuania
LY-AAA
YAK-40

Aurela

The Vilnius-based enterprise Aurela [–/LSK] bought a single Tu-134A, LY-ABD (ex-CCCP-665079, c/n [73]60054, f/n 4206) from Lithuanian Airlines in 1996, refitting it as a 43-seat Tu-134A 'Salon' registered LY-ASK. The aircraft remained in service until December 2000 when it was sold to Tulpar as RA-65079. *Yuriy Kirsanov*

In 2001, 2003 and 2004 Aurela purchased three Yak-42Ds in Russia, Kazakhstan and Cuba. LY-SKB bought from the Yakovlev Design Bureau (ex-RA-42362, c/n 4520424811431, f/n 0709) wore the basic colours of its previous owner (Elbrus Avia) and, before that, had seen service with Lithuanian Airlines as LY-AAX! In 2004, however, the trijets were sold to the Ukraine and Kazakhstan; LY-SKB is now UR-CDU with Tavrey Airlines. *Yefim Gordon archive*

Aviavilsa

The cargo carrier Aviavilsa [–/LVR] established in 1999 operated five An-26s from Vilnius over the years. The first aircraft, aptly registered LY-LVR, was leased from the Russian carrier Pskovavia in 1999-2002 and, like LY-FTL and LY-APC, was operated for DHL Worldwide Express. Here, LY-APN (ex-Ukraine Air Alliance UR-BXF, c/n 12010) shows off the airline's full livery while taxying at Helsinki; it is one of two currently in the fleet. *Eero Saarela*

Macedonia

META Aviotransport Macedonia

Meta Aviotransport Makedonija [M5/MAM] wet-leased two Tu-154Ms from Vnukovo Airlines in 1992. The first aircraft, RA-85745 *Ohrid* (c/n 92A928), is shown here; the other aircraft, RA-85619, was named *Ohrid II*. In August 1995 the airliners were returned to the lessor but retained this striking livery for a while. Meta Aviotransport Makedonija leased a pair of Yak-42s (UR-42376 and RA-42406) from Air Ukraine and Yugavia in 1993-94. *Yuriy Kirsanov*

Palair Macedonian

Seen here on short finals to Moscow-Sheremet'yevo, Tu-154B-1 LZ-BTJ (c/n 073411330, f/n 0902) is one of three Tu-154s wet-leased by Palair Macedonian [3D/PMK] from Balkan Bulgarian Airlines between 1991 and 1995. A further Tu-154 was leased from the Pulkovo Aviation Enterprise, flying from Skopje to Moscow and various destinations in Western Europe. *Yuriy Kirsanov*

Macedonian Airlines

MAT Macedonian Airlines (Makedonski Aviotransport) [IN/MAK] reportedly wet-leased Tu-154M LZ-BTQ from Balkan Bulgarian Airlines for a short while in 1995. Additionally, two Yak-42Ds, including RA-42326 (c/n 4520424402154, f/n 0406), were wet-leased from Saravia (Saratov Airlines) in the mid-1990s. *via Chris Knott*

Mali

Air Mali

When Mali proclaimed independence in 1960, deliveries of Western hardware were embargoed and the country sought help from the USSR. Among other things, Air Mali [L9/MLI] bought four IL-18s, including IL-18V TZ-ABE (ex-CCCP-75763, c/n 181003304). This aircraft was lost in a crash at Lingomin, Upper Volta, in 1974. IL-14s and An-2s were also operated. *Yefim Gordon archive*

An-24B TZ-ACK (c/n 87304202) was one of three delivered to Air Mali. It is now an instructional airframe at a technical school in Yegor'yevsk, Russia. *Yefim Gordon archive*

In 1996 Air Mali unexpectedly made a brief return to Soviet hardware, leasing Yak-42 LY-AAR (ex-CCCP-42345, c/n 4520422708304, f/n 0308) from Lithuanian Airlines in basic 1994-standard LAL colours complete with logos. *via Chris Knott*

Moldova

Air Moldova

Air Moldova [9U/MLD], the national flag carrier, was established in 1992 on the basis of the former Moldavian CAD's Kishinyov UAD comprising the 269th Flight (operating jets) and the 407th Flight (operating turboprops). The airline operated scheduled services to European destinations, as well as passenger charters to holiday spots and cargo charters. The original fleet included four An-24Bs, three An-24RVs, five An-26Bs, two An-32Bs, eight Tu-134As, three Tu-134AKs (one of which was used by the government flight) and eight Tu-154B/B-1/B-2s. The carrier's main base was at Kishinyov (Chişinău).

By 1999 Air Moldova had phased out the Tu-154, though the smaller Tu-134A remained in service. In 2001 the airline leased an Embraer 120 Brasilia turboprop and an Embraer ERJ145 jet (the latter has since been replaced by a second Brasilia). In 2003-04 it took delivery of two Airbus A320-211s used on flights to Europe. The rest of the current fleet comprises four An-24s, three An-26s, eight Tu-134A/AKs and one Yak-40D executive jet used for government VIP flights.

Originally Air Moldova was 100% state-owned. Now 49% of the shares have been acquired by a private investor, Unistar Ventures, but the government still has the controlling stock.

At first Air Moldova's aircraft wore nothing more decorative than basic Aeroflot colours with large red 'Air Moldova' titles and a white tail. Tu-134A-3 ER-65791 (c/n [03]63110, f/n 5602) is seen here landing at Moscow-Vnukovo in 1993. *Yuriy Kirsanov*

Sister ship ER-65036 (c/n 6348700, f/n 3703) illustrates the first full livery adopted by Air Moldova in 1994, which was not unlike the colours worn by Aeroflot's Tu-134As and Tu-134AKs prior to the introduction of the fleetwide standard livery in 1973. This aircraft is now a gate guard at Chişinău airport. *Yuriy Kirsanov*

Wearing Air Moldova's penultimate livery introduced in the spring of 2000, Tu-134AK ER-65140 (c/n [83]60932, f/n 4905) prepares to start its engines at Moscow/Vnukovo-1. It has now been transferred to the Moldovan government flight and repainted accordingly with 'Moldova' titles and the national coat-of-arms on the fin. *Yefim Gordon archive*

Tu-154B-2 ER-85332 (c/n 79A332) taxies in at Moscow/Vnukovo-1 after a flight from Kishinyov. The nondescript initial livery is augmented by 'Chartered by Balkan' additional titles dating back to a 1997 lease. The jet was sold to Chernomor-Soyuz in 1998, becoming RA-85332. *Yuriy Kirsanov*

Sister ship ER-85324 (c/n 79A324) shown here in Air Moldova's first full livery was also leased to Balkan – and also sold to Chernomor-Soyuz eventually as RA-85324. *Yuriy Kirsanov*

Still wearing the Soviet registration CCCP-46376 (c/n 07306009) but already adorned with Air Moldova titles and the Moldovan flag, one of Air Moldova's An-24Bs taxies towards the terminal at Kiev-Zhulyany. It became ER-46376 by 1994 and was retired by 2000. *Sergey Popsuyevich*

Air Moldova International

Air Moldova International [3R/MLV] was formed as a sister company of Air Moldova in 1995, operating a succession of aircraft leased in the Ukraine, Belorussia and Russia. These included An-24s, Yak-40s, such as Yak-40 'Salon 2nd Class' ER-88202 (ex/to EW-88202, c/n 9630949) leased from Belavia in 1995, and Yak-42s leased from Dniproavia and Bykovo Avia. The airline planned to introduce Embraer ERJ-145s in 2003 but went out of business in 2002. *via Martin Štepanek*

Moldavian Airlines

Moldavian Airlines [2M/MDV] were established on 26th July 1994 and started operations from Kishinyov on 19th August. This was the first private airline in Moldova. It became a Swiss-Moldovan joint venture in November 1999, forming a partnership with the Romanian airline Carpatair.

Most of Moldavian Airlines' fleet was leased from other carriers. Soviet hardware comprised five Yak-40s, one Tu-134A leased from Air Moldova and one Tu-134AK. At present Moldavian Airlines operate only Western aircraft.

Yak-40 'Salon 2nd Class' ER-JGD (ex-Samara Airlines RA-88307, c/n 9421334) wearing full Moldavian Airlines livery is seen here visiting Prague-Ruzyne. It was sold to GST-Aero as UN 87251 in 2004. *via Martin Štepanek*

Moldavian Airlines' fleet included this Tu-134AK, ER-TCF (ex-Vichi Aircompany ER-AAZ, c/n [93]62390, f/n 5301) – a former Moldovan Air Force Tu-134 Balkany airborne command post. Note the fairing under the APU jetpipe where an aft-pointing aerial used to be. Tu-134A-3 ER-65036 was also operated briefly. *Yuriy Kirsanov*

Aeriantur-M

Aeriantur-M [–/TUM] **was established in 1994, operating passenger and cargo charters with An-72/An-74 freighters, Tu-134s and Tu-154s leased as required. An-26B-100 ER-26068 (c/n 11308) was converted from a standard freighter acquired from Air Moldova in 2002; the extra cabin windows and the large emergency exit to port are visible here. The only other aircraft in the fleet is an An-12 bought from the Russian carrier Avial'-NV.** *Martin Novak*

Tepavia Trans

Tepavia Trans [–TET] founded in 1999 started out with a fleet consisting of eleven An-28 twin-turboprop feederliners. Most of them wore the blue/white livery exemplified by ER-AJC (ex-RA-28748, c/n 1AJ 003-12), although some were anonymous. As of this writing, part of the original An-28 fleet has been sold or replaced; however, Tepavia Trans has also added two An-32 freighters and a pair of An-12s to its inventory. *Sergey Popsuyevich*

Moldavian Government

The Moldovan government flight operated a succession of VIP jets which included Tu-134AK ER-65094 (later superseded by ER-65140) and Yak-42D 'Salon' ER-YCA. Here, Yak-40 ER-JGD (see previous page!) is seen starting up at Prague; oddly, the aircraft does not appear to be fitted with a VIP interior.
via Bob Archer

Mongolia

Mongolian Airlines

Mongolian International Air Transport (MIAT), now known as Mongolian Airlines [OM/MGL], had operated Soviet airliners since 1956. Aircraft in MIAT service included the IL-14 (11 delivered), An-2 (more than 80 delivered), An-24 (20 delivered), which was the mainstay of Mongolia's civil aviation since the early 1960s, An-26 (four) and An-30 (one). However, bigger and more sophisticated aircraft were required for international services, so MIAT leased a Tu-154B-2 from Aeroflot in 1988-1990. It was superseded by this Tu-154M, MPR-85644 (ex/to RA-85644, c/n 88A780), which is shown here landing at Moscow-Sheremet'yevo. *Sergey Sergeyev*

Mozambique

Linhas Aereas de Moçambique

The only known Soviet aircraft operated by the national airline LAM (Linhas Aereas de Moçambique) were an An-26 freighter and this IL-62M (C9-BAE, c/n 3344724) delivered in 1984 used as a government VIP transport. It is shown here in its ultimate livery. C9-BAE was flown to Russia for refurbishment in 1991, but Mozambique could not pay for the work and abandoned the aircraft, which was subsequently sold to the Russian airline Nebo as RA-86931. *Yefim Gordon archive*

Nigeria

Harka Air

In mid-1994 **Harka Air**, an airline based at Lagos-Mohammed Murtala, wet-leased two Tu-134As and six Tu-134AKs from the Russian carrier Komiavia. Additionally, the carrier leased three Yak-40s from Mogilyov-Avia, one from Smolensk-Avia and one or two from Air Lithuania. The Russian jets, which were operated on domestic services, were popular with the passengers but Harka Air proved reluctant to pay the lessors and the leases were terminated in 1995.

Seen here taxying at Moscow-Sheremet'yevo in 1995, radar-nosed Tu-134AK RA 65609 (c/n [63]46155, f/n 3401) of Komiavia still wears the titles and logo of recent lessor Harka Air in combination with the basic colours of the machine's first owner, the East German airline Interflug. *Dmitriy Petrochenko*

Apart from Russia, Harka Air leased aircraft from Lithuania. LY-AAC (c/n 9530344) is one of the two leased from Lietuva (Air Lithuania) in basic old Lietuva colours; before that, it was operated by Nigeria Mails. *via Chris Knott*

Premium Air Shuttle

Premium Air Shuttle (PAS) was another Nigerian Yak-40 operator, with up to five of the type in the fleet since 1997. The aircraft wore very different colour schemes; 5N-BOS (ex-LZ-DOA, c/n 9341431) looked very smart. Originally delivered to Balkan Bulgarian Airlines and then sold to Hemus Air, the aircraft had served with several African carriers before being placed on the Nigerian register. By 2006 two of the carrier's Yak-40s had apparently been sold to the CIS. *via Chris Knott*

Kolkol Airlines

Seen here arriving at Moscow/Vnukovo-1 on the daily Air Moldova service from Kishinyov, Tu-134A ER-65051 (c/n [73]49758, f/n 3810) still carries Kolkol Airlines titles from a lease that never was – the Nigerian carrier planned to lease this aircraft and two other Tu-134As in 1994 but proved insolvent. *via Martin Štepanek*

Nicaragua

Aeronica

Aeronica – Aerolineas Nicaragüenses [RL/ANI] based at Managua was the Nicaraguan flag carrier established in 1981 after the Sandinista revolution. It operated domestic routes and international services to Guatemala, El Salvador, Honduras, Costa Rica, Panama, Cuba, Mexico and the United States. The fleet included two An-26s, two An-32s and a single Tu-154M delivered in December 1989. This was reportedly the presidential jet of Daniel Ortega Saavedra, the leader of the Sandinista party.

In 1992, when the civil war in Nicaragua ended, the Sandinistas were removed from power. A wave of change swept the country; one of the results was the demise of Aeronica.

An-26 YN-BYW (c/n 12404) in Aeronica's old colours stages through Prague-Ruzyne on its delivery flight on 18th September 1988; note the registration repeated in smaller characters below the flag on the tail. *Martin Novak*

An-32 YN-CBV (c/n 1809) shows Aeronica's new livery. This aircraft was sold or leased to the Colombian airline TAP as HK-4132X in 2000 but reverted to YN-CBV in 2002. *Martin Novak*

North Korea

Air Koryo

Established in 1954 as Chosonminhang (CAAK – Civil Aviation Administration of Korea) [JS/KOR], the airline started operations on 21st September 1955 as the successor of SOKAO, the joint Soviet-North Korean airline established in 1950. CAAK began operations with Li-2s and An-2s. At least five IL-14 and four IL-18 turboprops were added to the fleet in the 1960s; from 1966 onwards the IL-14s were replaced by eight An-24s.

Jet operation commenced in 1975, when the first of four Tu-154Bs/B-2s was delivered for services from Pyongyang to Prague, East Berlin and Moscow. However, since the Tu-154 lacked the range for non-stop services to Moscow (to say nothing of destinations further west), CAAK ordered the IL-62M; the first of an eventual six was delivered in 1982. In 1984 the jet fleet was augmented by a pair of Tu-134B-3s. The North Korean Air Force's three IL-76MDs were also used by the airline. CAAK also functioned as the government VIP flight.

The end of the Cold War and the collapse of socialism in Europe saw a reduction in the number of international services. CAAK was renamed **Air Koryo** in 1993. No new aircraft have been purchased since 1990 and the current fleet is elderly, although an order for two Tu-204-300s has been placed recently. As of 22nd March 2006, Air Koryo was banned from EU airspace due to unsatisfactory operating standards, although North Korea's internationally denounced efforts to create nuclear weapons were probably a contributing factor.

IL-18D P-835 (c/n 188011205), one of the four IL-18s delivered to North Korea, in the old Chosonminhang (CAAK) colours. *Martin Novak*

As this photo of IL-62M 'Salon' P-618 (c/n 2546624) arriving at Moscow-Sheremet'yevo testifies, Chosonminhang aircraft were sometimes operated without titles. After all, everybody knows there is only one airline in North Korea... or perhaps the reason is that this is a government VIP aircraft. *Yuriy Kirsanov*

Again photographed at Moscow-Sheremet'yevo, IL-62M P-885 (c/n 3933913) shows off the new titles and logo of Air Koryo. The grey-painted tailplane fairing is noteworthy, as is the extreme negative tailplane incidence typical of the IL-62 in landing configuration. *Yuriy Kirsanov*

Opposite page:

IL-76MD 'Falsie' P-913 (c/n 1003404126, f/n 7902) in its original colour scheme. Interestingly, the designation is not painted on the aircraft. *Martin Novak*

Caught by the camera a few seconds before touchdown, the same aircraft is seen here in an interim 'hybrid' livery with Air Koryo titles but still with a large North Korean flag on the tail instead of a logo plus a small logo on the nose. Once again, note that the tailplane bullet fairing is now painted grey. *Yuriy Kirsanov*

The airline's third IL-76MD 'Falsie', P-914 (c/n 1003404146, f/n 7907), painted in Air Koryo colours 'by the book'. *Yuriy Kirsanov*

P-913
조선민항

P-913
AIR KORYO
고려항공

P-912
AIR KORYO
고려항공

Chosonminhang's third Tu-154A, P-553 (c/n 77A191), is seen during a visit to Prague. *Martin Novak*

Tu-134B-3 P-813 (c/n [43]66215, f/n 6354) in hybrid Air Koryo livery (with new titles but old tail colours) on final approach. Despite the seeming presence of a white 'plug' in the first full-size window to starboard (indicative of a forward toilet and hence of a new-build VIP version), this is not a Tu-134B-3 'Salon' similar to the Tu-134AK: P-813 has no rear entry door and airstairs. *Yuriy Kirsanov*

Pakistan

Bhoja Air

Founded in 1993, Karachi-based Bhoja Air [B4/BHO] leased Yak-42s from Lithuanian Airlines as required until 2000. Four aircraft were leased in this fashion; LY-AAV (ex-CCCP-42355, c/n 4520424711399, f/n 1308) is shown here; note the small Pakistani flag aft of the 'Yak-42D' nose titles. *Sergey and Dmitriy Komissarov collection*

Shaheen Air International

Shaheen Air International [NL/SAI], also based at Karachi, operates a mix of airliners leased from East and West alike. These included seven Tu-154s leased in Belorussia and Russia between 2001 and 2005 and at least one Yak-42. Here, Tu-154B-2 EW-85538 (c/n 82A538) is seen at Minsk-2 on 20th May 2005 after the Pakistani lease; the basic colours are those of Belavia. The airline also operated Yak-42s leased from the Ukrainian carrier Dniproavia. *via Peter Davison*

Poland

LOT

LOT Polish Airlines [LO/LOT] (Polskie Linie Lotnicze LOT; 'lot' is Polish for 'flight') was established on 1st January 1929 as a state-owned corporation operating Dutch, German and US types. Services were suspended during the Second World War, resuming in 1945, with 39 Li-2Ps and Li-2Ts (many of them transferred from the Soviet Air Force) as the first post-war equipment.

Even though Poland was now within the Soviet sphere of influence, LOT continued operating airliners from East and West alike for a while. Five SNCASE SE.161 Languedocs joined the fleet in July 1947, followed by five IL-12Bs in 1949 and six IL-14Ps in 1955 (augmented by Polish Air Force IL-14TDs leased for seasonal cargo services from 1961 onwards). The Convair 240 and Vickers Viscount were first acquired in October 1957 and November 1962. Later, however, the composition of the fleet shifted to Soviet aircraft. The IL-18 was introduced in May 1961, allowing the establishment of routes to Africa and the Middle East; six IL-18Vs and four 'Es were delivered. 20 An-24Bs delivered from April 1966 replaced the IL-14s, followed by five Tu-134s and five Tu-134As in 1968-76. Seven IL-62 long-haul airliners delivered in 1972-77, replaced by nine IL-62Ms in 1979-87, enabled transatlantic services to Montreal and New York. 14 Tu-154Ms delivered between May 1986 and January 1991 replaced the IL-18s.

In 1989 the Polish flag carrier started to re-equip with Western hardware, replacing the IL-62Ms with Boeing 767s and the Tupolev fleet with Boeing 737s and Embraer Regional Jets. By mid-1995 all of LOT's Soviet aircraft had been grounded and eventually sold or scrapped.

The advent of the twinjet Tu-134 caused the IL-18 turboprops to be relegated to the freighter role. Here, IL-18V SP-LSD (c/n 184007102) is seen in LOT's 1977-standard colours with additional 'Cargo' titles. *Yefim Gordon archive*

The fifth and last 'glass-nosed' Tu-134A delivered to LOT, SP-LHE (c/n [63]48405, f/n 3507), demonstrates the pre-1977 livery worn by the type. *Sergey and Dmitriy Komissarov collection*

The same aircraft parked at a Polish airport in 1977-standard colours, with an SPT-154 self-propelled boarding ramp parked alongside. The aircraft was retired in 1993 and is now part of an aviation and military hardware display at Łódź-Lublinek. *Sergey and Dmitriy Komissarov collection*

IL-62 ***sans suffixe*** SP-LAB (c/n 21105) in pre-1977 colours taxies at Warsaw-Okęcie. The aircraft is named ***Mikolaj Kopernik*** (Nicholas Copernicus), and the scientist's portrait is visible near the forward entry door. SP-LAB was sold to Aeroflot as CCCP-86706 in 1983 and went on to serve with Kras Air as RA-86706. *Yefim Gordon archive*

SP-LBD ***General Wladyslaw Sikorski*** (c/n 1138234), the carrier's fourth IL-62M. In 1992 the aircraft was sold to Air Ukraine, which operated it as CCCP-86133 and subsequently as UR-86133 in basic LOT colours. *Yuriy Kirsanov*

Tu-154M SP-LCO (c/n 90A862) starts up the engines at Warsaw-Okęcie. This machine is one of the two that were not re-exported to the CIS, being transferred to the Polish Air Force in September 1994 and refitted as Tu-154M 'Salon' '862 Black' (the current serial is '02 Red').
via Martin Štepanek

An-24B SP-LTB *Bug* (pronounced 'Boog' – the name of a river; c/n 67302205) in 1977-standard livery sits with a power cart hooked up and ready for engine starting. This aircraft, too, was re-exported to the Ukraine, and became UR-49248 with Air Ukraine in 1992.
Yefim Gordon archive

The Yak-40s of the Polish Air Force's, government flight (the 36th Special Transport Regiment), including Yak-40 'Salon 2nd Class' SP-LEE (ex-'047 Red', c/n 9021560), were leased to LOT in 1988. The aircraft reverted to its military identity as '047 White' after the lease, wearing Polish Air Force insignia over basic LOT colours until repainted in the current red/white scheme as '047 Red'.
via Martin Štepanek

LOT also leased Polish Air Force An-26s for cargo/mail flights from time to time, sometimes repainting them in full airline colours. SP-LWB (ex/to '1603 Black', c/n 1603) was leased from the transport regiment at Krakow-Balicein from 1989 to 1996. *Yefim Gordon archive*

Another example of Polish Air Force transports being temporarily operated by LOT is An-12BP SP-LZB (ex-'51 Red', c/n 6344308). It was leased as such on several occasions and eventually sold to Air Sofia in 1995 as LZ-SFS. *Wacław Hołyś*

Exin

Founded in 1991, the cargo airline Exin [–/EXN] operates an all-An-26 fleet from Katowice. Here, SP-FTZ (ex-Belavia EW-26632, c/n 5909) is seen in non-airworthy condition at Minsk-1 in modified DHL colours. This aircraft was reportedly sold to Lithuania in 1999 but the picture was taken in May 2005. *via Peter Davison*

Exin's An-26s wore a variety of colour schemes, as illustrated here by An-26B SP-FDT (ex-Altai Airlines RA-26116, c/n 27312205) bought in 2001. *Yefim Gordon archive*

Exin is a long-standing partner of DHL Worldwide Express, and some of its aircraft, including An-26B SP-FTL (ex-Belavia EW-26616, c/n 5002), wore full DHL colours. This aircraft was sold to Aviavilsa in 19999 as LY-FTL. Ten An-26s have been operated over the years. *Sergey and Dmitriy Komissarov collection*

Romania

TAROM

Established on 8th August 1945 after Romania fell under Soviet influence, the Soviet-Romanian airline **TARS** (Transporturile Aeriene Romåno-Sovietice) was the successor of the pre-war airline which had operated since 1920 as CFRNA, since 1926 as CIDNA and since 1930 as LARES. Domestic operations were started from Bucharest-Baneasa airport on 1st February 1946 with Li-2Ps, of which 26 were delivered. The Romanian government later bought the company's Soviet share and, on 18th September 1954, the airline was renamed **TAROM [RO/ROT]** (Transporturile Aeriene Române – Romanian Air Transport).

Like most of the airlines belonging to Soviet Bloc states, TAROM operated primarily Soviet-designed aircraft. The Li-2s were supplemented in 1956-59 by 19 IL-14Ps, which allowed the route network to be expanded; by 1960, the airline was flying to a dozen cities across Europe. In 1961 TAROM received its first turbine-powered aircraft, purchasing nine IL-18Vs, two IL-18Es and one IL-18D by 1967, which made long-haul services possible; 1966 saw the operation of its first trans-Atlantic flight. Three IL-62s delivered in 1973-75 and two IL-62Ms delivered in 1977-78 took over some of the routes; the medium-haul international services were performed by five Tu-154Bs, three Tu-154B-1s and four Tu-154B-2s, while six An-24Bs and 30 An-24RVs replaced the piston-engined airliners. In parallel, TAROM operated Western aircraft, being an exception to the rule; six Boeing 707-320s were acquired in 1974 for use on the prestigious intercontinental routes, while short-haul services were performed by 17 BAC One-Elevens. It was only in the wake of the collapse of the Soviet Union in 1991 and the fall of the Ceausescu regime in Romania that the airline was finally able to acquire Western-built jets of its own choice. By the turn of the century all Soviet-built aircraft had been withdrawn.

Seen visiting Prague-Ruzyne, An-24RV YR-BMN (c/n 77310808) illustrates the early TAROM livery. Note the non-standard tapered blade aerial aft of the nose gear unit. *Martin Novak*

The same aircraft in the Romanian flag carrier's current livery. YR-BMN was sold to Aerovista in 2001 along with several sister ships, becoming EX-24808 and later 4L-AVL. *Yefim Gordon archive*

Tu-154B-1 YR-TPF (c/n 77A239) in old TAROM colours awaits delivery at Kuibyshev-Bezymyanka, the factory airfield of plant No 18 which built it. Like most of the carrier's Tu-154s, it was eventually retired and scrapped in post-Communist days.
Yefim Gordon archive

Here the same aircraft retracts its undercarriage after take-off from Moscow-Sheremet'yevo, displaying the late 1980s livery with red titles and a red logo on a white tail.
Yuriy Kirsanov

IL-62M YR-IRD (c/n 4727546), the first of two delivered to TAROM, on short finals to Moscow-Sheremet'yevo in its second livery. The carrier also operated three IL-62s *sans suffixe* (YR-IRA through YR-IRC). *Yuriy Kirsanov*

The same aircraft in its ultimate livery in Romanian service. Together with the other IL-62M and two of the three IL-62s *sans suffixe* it was sold to Kazakh carrier Trans Asian Airlines in 1998 but was written off in an accident at Istanbul-Ataturk International on 24th April that year before it could receive a Kazakh registration. *Yuriy Kirsanov*

São Tomé & Principe

Express International

'IL-76T' (actually a former IL-76K zero-G trainer) 3C-QQD of Express International Cargo Corp. (ex-S9-BOM, ex-RA76372, c/n 073410279, f/n 0710) is pictured at Sharjah in late 2000. *Sergey and Dmitriy Komissarov collection*

Many of the commercial aircraft registered in Sao Tome & Principe are 'anonymous', with no indication of the operator. Seen here at Sharjah in 2004, IL-76MD S9-DAE (ex-Atlant Hungary HA-TCH, ex-Khors UR-76438, ex-CCCP-78751, c/n 0083483513, f/n 6309) had been operated by the Belgian carrier Skylink earlier that year but wore no titles when the picture was taken. *Aidan Curley*

Another anonymous visitor to Sharjah, An-72 S9-GRE (c/n unknown) wore the basic colours of Centrafrican Airlines in 2000. *Sergey Popsuyevich*

Serbia & Montenegro

Kosmas Air

Established in 2004, Belgrade-based Kosmas Air (the name means 'space' in Serbian) operates a single aircraft, IL-76TD YU-AMI (ex-RA-76822, c/n 0093499982, f/n 7506). The basic livery and tail logo are those of ex-owner Titan Cargo. *Yefim Gordon archive*

Slovakia

Univerz Air

Univerz Air operated two An-2 biplanes in the sprayer and utility role. One of them, OM-UIN (ex-OK-UIN, c/n 1G233-43) was operated for the British company Avia Special based at White Waltham, which used the aircraft for paradropping and joyriding. The machine is seen here attending an airshow in the UK in the early 1990s. In 1996 OM-UIN was reregistered OM-248 and then sold to Avia Special, being placed on the Hungarian register as HA-MKF for the sake of convenience. *Yefim Gordon archive*

Air Transport Europe

Air Transport Europe [–/EAT], primarily a helicopter operator based at Poprad, bought Tu-134A ES-AAF (ex-CCCP-65034, c/n 6348565, f/n 3701) from Estonian Air in 1995. The jet was operated as OM-GAT *David* until November 2000, when it was sold to the Moldovan carrier Tiramavia as ER-TCH. *via Martin Štepanek*

Slovakian Airlines

Bratislava-based **Slovakian Airlines/Slovenské Aérolinie [6Q/SLL]** took delivery of three Tu-154-100s in 1998 as part payment of Russia's state debt to Slovakia. They were equipped with American avionics and French cabin interiors. In addition to using the trijets on its own services, Slovakian Airlines occasionally leased them abroad. In 2003 the carrier replaced these aircraft with Boeings, and the Tu-154-100s were sold to Russia at the end of the year.

Tu-154-100 OM-AAB (c/n 98A1015) taxies out for take-off out at Moscow/Sheremet'yevo-2, showing the additional titles to the effect that Bratislava was a candidate to host the 2006 Winter Olympic Games. In late 2003 the carrier sold all three aircraft to the Pulkovo Air Enterprise; OM-AAB became RA-85835. *Yuriy Kirsanov*

Slovakian Government

Still wearing its original Czech registration OK-BYO but soon to become OM-BYO, this Tu-154M 'Salon' (c/n 89A803) has a red/white livery reminiscent of the one worn by the Czech federal government flight Tu-154B-2 'Salons' in the 1980s. *Yefim Gordon archive*

By comparison, Tu-154M 'Salon' OM-BYR (c/n 98A1012) delivered in March 1998 as part payment of Russia's state debt to Slovakia wears a predominantly white colour scheme. *Martin Novak*

Another Slovakian government jet, Yak-40 OM-BYE (ex-OK-BYE, c/n 9440338), in the same white livery with Slovak flag stripes and 'Slovenska Republika' titles. The aircraft has apparently been refitted from a Yak-40 'Salon 2nd Class' to a regular passenger aircraft, being regularly operated for Tatra Air. *via Martin Štepanek*

Slovenia

Falcon Air

Falcon Air [–/FAR] based at Lesce operates three An-2 utility biplanes, including S5-CAR (ex-CCCP-54903, c/n 1G186-51) purchased in November 1994. The aircraft is aptly named *Annushka* (Little Annie – the traditional nickname of the An-2). *Yefim Gordon archive*

Somalia

Air Somalia

Air Somalia wet-leased a single Tu-154B-2, UR-85546 (c/n 82A546), from Tavria-MAK in December 2000. During the one-year lease the aircraft, which operated from Sharjah, changed its identity to ER-TAI. In January 2002 Air Somalia suspended operations and the aircraft went to Air Service International. *via Bob Archer*

South Africa

Romoco Cargo

Romoco Cargo SA operated An-74 UR-74026 (c/n 365.470.96.919, f/n 1506) from the aero engine manufacturer Motor Sich in 1995-96; it is seen here at Lanseria, South Africa, where it was based for the duration. Note the international civil aid and protection symbol on the tail. *Peter Davison*

Sri Lanka

ExpoAir

Expo Aviation [8D/EXV] (aka ExpoAir) operated two An-12s, including civil-configured An-12BP 4R-EXC (ex-RA-12956, c/n 8345507) purchased from Samara Airlines in 1999; it is seen here being resprayed in the open at Sharjah. The airline also has a single IL-18D. *Aidan Curley*

Bismillah Airlines

Bismillah Airlines [5Z/BML] were established in 1998. The airline, which is owned by the Mollah Group of Industries, is the first private airline licensed to operate regular international cargo services as a Bangladeshi carrier. The fleet includes two Cambodian-registered An-12s; a single An-8 registered in Equatorial Guinea (3C-ZZO, c/n 0D 3440) was also operated in 2000-01. *Yefim Gordon archive*

Azza Transport

Azza Transport Co Ltd [–/AZZ] was founded in 1993, operating a cargo charter service throughout Africa and the Middle East, with planned services to Europe. Its main base is Khartoum; an auxiliary hub has now been set up at Sharjah, UAE. The carrier started operations with two Boeing 707-320Cs, but its first IL-76TD was purchased from Russia in 1996. A second (ex-Iranian) IL-76TD was added in 2000 and an ex-Moldovan An-12BK in 2002; the carrier also reportedly has a single An-26. Additionally, AZZA allegedly operated a single Tu-134A airliner.

Azza Transport IL-76TD ST-APS (ex-RA-76837, c/n 1023409316, f/n 8309) taxies at Sharjah in 1997. It is still in the basic red/white colours of the Belorussian airline Belair with which it had served as EW-76837. *Sergey Popsuyevich*

The same aircraft after gaining the full livery of Azza Air Transport except for the missing titles. *Yefim Gordon archive*

Ababeel Aviation

Ababeel Aviation is one of the many small Sudanese air carriers that come and go almost faster than you can keep track of them. The airline operated at least two An-24s, four An-26s and two IL-76TDs. An-26 ST-AWT (ex-Kit Air RA-58646, c/n 3508) is shown here at Sharjah in February 2004; this aircraft was previously operated by Air West Express. *Aidan Curley*

East/West Cargo

Established in 2003 as the sister company of the Sudanese carrier Air West [–/AWZ], East/West cargo operated at least three IL-76TD ‘Falsies’ under the same flight code. ST-EWC (ex-Botir Avia EX-86919, c/n 0023438129, f/n 2903) is seen here on short finals to Sharjah. The airline’s operating licence was revoked after ST-EWB crashed near Khartoum in February 2005; ST-EWC is now operated by Air West. *Martin Novak*

Juba Air Cargo

Named after a city in southern Sudan, Juba Air Cargo [–/JUC] has been wet-leasing An-12s, An-26s, An-32s and IL-76s as required since 1996. Most of the aircraft were/are Sudanese-registered, but IL-76TD ER-IBL (ex-Atruvera Airlines RA-76659, c/n 0053463908, f/n 4807) leased from Airline Transport, Inc. in 2005 was on the Moldovan register. The aircraft used to be an IL-76TD ‘Falsie’ but has had the tail gunner’s station replaced with a scratchbuilt commercial tailcone, becoming a ‘true’ TD. It is now registered EX-075 with an unknown operator. *Aidan Curley*

Sasco Air Lines

Sasco Air Lines (Sudanese Aeronautical Services Co) [–/SAC] lease passenger and cargo aircraft from CIS airlines as required. Currently Sasco operates a single An-24RV leased from the Ukrainian carrier South Airlines.

Aptly named *Cinderella*, Kyrghyz-registered An-12BP EX-034 (ex-Imtrec Aviation XU-365 No 1, ex-UK-11109, c/n 01348002) pictured here at Sharjah in late 2004 sported the full colours of Sasco Air Lines. It is currently in service with a different outfit called TCS. *Aidan Curley*

Tu-134A-3 EY-65876 (c/n [53]31220, f/n 3002) takes off at Moscow-Vnukovo. The 'Sasco Air Lines' titles are in small type on the tail. This aircraft was operated in 1995-98. *Yuriy Kirsanov*

Sudan Airways

The nation's flag carrier Sudan Airways [SD/SUD] occasionally leases Antonov and Yakovlev aircraft. This An-24RV (c/n 67310505) was leased as RA-47318 in 2002 and became EK-47318 during the lease. It wore basic Sudan Airways livery but the colours are wrong, the cheatline and tail being orange and dark blue instead of yellow and black. It is now in service with Daallo Airlines as J2-SHE. An-24T EK-49273 wore the correct colours. *Aidan Curley*

Forner Airlines

Forner Airlines operated at least two civil-configured An-12BPs, including ST-AQQ (ex-Mikma ER-ACG, c/n 9346504); the blue/white colours were those of National Commuter Airlines that had operated the aircraft as D2-FCT in 2001. The freighter is seen here at Sharjah in August 2002 with additional SSA (Sudanese State Aviation) titles. The latter carrier bought this aircraft in 2003. *Aidan Curley*

Sarit Airlines

Founded in 1998, Sarit Airlines [–/SRW] operated Antonov propliners from Khartoum. Aircraft that saw service with the carrier included two An-26Bs and seven An-12s, including an Azeri-registered example on lease from Silk Way Airlines. An-12B ST-SAR (ex-Lasare Air 4L-CAA, c/n 402102) illustrates the carrier's smart livery at Sharjah in early 2003; it crashed at Wau, Sudan, on 24th September that year. An IL-76TD was leased from Georgia in 2005. The airline is now known as Badr Airlines [–/BDR], operating single examples of the An-26, An-28, An-72 and An-74. *Yefim Gordon archive*

Sudan Government

Immaculate-looking An-74 ST-PRB (ex-Proton RA-74024, c/n 365.470.96.918, f/n 1505) in VIP/combi configuration awaits delivery to the Sudanese government flight at Moscow-Ostaf'yevo in 2005. The Sudanese government also makes use of new-build IL-62M 'Salon' ST-PRA. *Sergey Krivchikov*

Data Airlines

Data Airlines or Data International Aviation [–/DTN] had two An-12s registered ST-APJ and ST-APU. The former aircraft, a Voronezh-built An-12A (ex-Avkom RA-11308, c/n 2400701), is seen here undergoing maintenance at Sharjah in March 2002; the titles on the port side are in Arabic. A third An-12 (ST-APG) was also reported – possibly in error. *Dmitriy Petrochenko*

Swaziland

Air Pass

Air Pass based at Pietersburg-Gateway (South Africa), a sister company of the Liberian carrier Air Cess, took over the latter's fleet in 1997. The fleet included three An-12s, one An-24RT, three An-32s, two An-72s, five IL-18s, three IL-76s, one Tu-154M and four Yak-40s. In 1998, however, Air Pass suspended operations, transferring almost its entire fleet to Centrafrican Airlines. Here, anonymous-looking An-32 3D-RTD (ex-RA-48109, c/n 1708) takes off from Sharjah. *Sergey Popsuyevich*

Syria

Syrianair

Syrianair (Syrian Arab Airlines) **[RB/SYR]**, the sole national airline, traditionally operated Western types (primarily Boeings). However, since Syria was on good terms with the Soviet Union and an active recipient of Soviet military aid, Syrianair was obliged to order Soviet aircraft as well. Thus in 1985 the airline became the second export customer for the Tu-154M, taking delivery of three early-production aircraft in March-June. They were used on the Damascus-Moscow service and flew to various other destinations in Europe (including Athens) and the Middle East. Another Tupolev aircraft operated by Syrianair was the Tu-134B-3. Three of the six examples delivered between 1982 and 1984 were built in VIP configuration similar to the Tu-134AK and used for transporting high-ranking statesmen, although President Hafez Assad still preferred using a Boeing 727 or a Dassault Falcon 50.

In 1998 Syrianair began a large-scale fleet renewal programme. The Tupolevs were replaced by six new Airbus Industrie A320-232s; by March 2001 the Soviet jets had been placed in storage at Damascus-International. Yet, Russia has not lost the Syrian commercial aviation market completely: talks are under way concerning the purchase of IL-96 long-haul airliners and Tu-204 medium-haul twinjets by Syrianair.

Additionally, the Syrian Air Force's four IL-76T freighters wear Syrianair colours and are periodically used by the airline on commercial flights.

Syrianair's third and final Tu-154M, YK-AIC (c/n 85A710), lines up for take-off from runway 25L at Moscow-Sheremet'yevo. *Dmitriy Petrochenko*

Three of Syrianair's Tu-134B-3s, including YK-AYD (c/n [23]63990, f/n 6328), were in VIP configuration. The aircraft is shown here at Prague-Ruzyne in its original livery. *Martin Novak*

Pictured here taxying at Moscow-Sheremet'yevo, another VIP-configured Tu-134B-3 (YK-AYB, c/n [23]63994, f/n 6331) shows off Syrianair's current colours adopted in 1998. Note the rear entry door characteristic of new-build VIP examples. *Sergey Sergeyev*

SYRIANAIR
YK-AIC

السورية
SYRIANAIR
Tu-134B-3
YK-AYD

АЭРОФЛОТ-ДОН
السورية
YK-AYB
SYRIAN

The Syrian Air Force's IL-76, which were frequent visitors to Moscow-Sheremet'yevo, wore Syrianair colours or titles and were used by the airline from time to time. YK-ATA (c/n 093421613, f/n 1604) shown here and YK-ATB were delivered as 'true' IL-76Ms in this livery typical of Syrian quasi-civil aircraft. Both have now been converted to 'true' IL-76Ts with solid tailcones and repainted in the current colours. *Yuriy Kirsanov*

In contrast, the other two aircraft, YK-ATC and YK-ATD (c/n 0013431915, f/n 2309), were IL-76MD 'Falsies' with no tail gunner's station and wore full Syrianair colours. They, too, have now been refitted as IL-76Ts. *Yefim Gordon archive*

Tajikistan

Tajikistan Airlines

Tajikistan Airlines [W5/TZK, later 7J/TZK], formerly **Tajik Air**, was established in 1990 as the successor of Aeroflot's Tajik CAD. Tajikistan Airlines performs scheduled services within the CIS and to Munich, New Delhi, Sharjah and Urumqi, as well as charters. The airline's main bases are at Dushanbe and Khujand (formerly Leninabad), with hubs at Kulyab and Khorog. The current fleet includes three An-28s (of the initial 12), seven Yak-40s (of the initial 20), one Mi-8T (of the initial eight) and four Mi-8MTV-1s operated by the former Dushanbe UAD/178th Flight, five Tu-154B-2s (of the original nine) and four Tu-154Ms flown by the Dushanbe UAD/186th Flight, three An-24Bs, three An-24RVs, two An-26s and seven (of the initial 12) Tu-134A-3s operated by the former Leninabad UAD/292nd Flight. The three Tu-154B-1s are now retired.

Tu-154B-2 EY-85511 (c/n 81A511) wearing full Tajikistan Airlines livery vacates runway 14L at Moscow-Domodedovo after arriving from Ashgabat. *Yuriy Kirsanov*

Tajikistan Airlines' fleet includes four Tu-154Ms, exemplified by EY-85691 (c/n 90A864) seen here 'cleaning up' on departure from Moscow. This aircraft has been leased to foreign operators many times. *Yuriy Kirsanov*

The forward entry door of Tu-154M EY-85692 (c/n 90A865) carries the national coat-of-arms, although it is not a government VIP jet. *Dmitriy Komissarov*

Initially many of the carrier's aircraft, such as Tu-134A-3 EY-65788 (c/n [03]62835, f/n 5509), wore basic Aeroflot colours with a Tajikistan Airlines logo but no titles. *Yuriy Kirsanov*

Seen here during a quick turnaround at Moscow-Domodedovo on a bleak snowy day in late November 1998, Tu-134A-3 EY-65022 (c/n [63]48395, f/n 3505) is a re-export aircraft from Estonia, having served with Estonian Air until January 1997 as ES-AAE. Tu-134A EY-65127 was formerly ES-AAJ with the same carrier but presumably never entered service, being sold to Russia as RA-65127. *Dmitriy Komissarov*

Pictured at Sharjah in the late 1990s, An-26B EY-26205 (c/n 14107 is one of two currently in the fleet. Some of the aircraft have titles in the Tajik language. *Yefim Gordon archive*

Turkey

CAT Cargo

CAT Cargo was one of two Turkish operators of the An-12, leasing An-12B RA-11366 (c/n 402808) from Magadan Cargo Airlines in 1994. Reregistered TC-KET, the aircraft was operated until 1997 in this simple but pleasing livery. The aircraft was still in storage at Istanbul-Ataturk International in 2004. *Dmitriy Petrochenko*

Greenair

Established in 1990, the Istanbul-based charter carrier **Greenair [–/GRN]** (a division of Sen Holding) operated three 164-seat Tu-154Ms and a 350-seat IL-86 wet-leased from Vnukovo Airlines, as well as two Tu-134AKs leased from the Rossiya State Transport Co (the Russian government flight). The jets were used for carrying tourists from all over Europe to Turkish holiday resorts. The airline ceased operations in 1995, returning the aircraft to their respective owners.

Tu-154M TC-GRA *Cappadocia* (ex-RA-85620, c/n 87A739) is towed to the terminal. The carrier's other two Tu-154Ms, TC-GRB and TC-GRC, had a slightly different livery with a bigger dark green area on the rear fuselage. *RART*

Holiday Airlines

The charter carrier Holiday Air [HW/HLD] briefly operated two Tu-154Ms, including TC-RAD (ex/to RA-85714, c/n 91A890) leased from ALAK, in 1995-96; the aircraft is seen here taxying at Moscow/Sheremet'yevo-2 in August 1995. Additionally, a single Yak-42D (UR-42540) leased from Air Ukraine was operated within the same time frame. *Dmitriy Petrochenko*

Mas Air

Istanbul-based Mas Air which, like Greenair, was associated with Sen Holding, operated three An-24RVs leased in Russia and the Ukraine. One of them, TC-HSA *Bodrum* (ex/to Bykovo Avia RA-46504, c/n 37308308), is seen here at Moscow-Bykovo after returning from the lease. Mas Air also operated at least three Mi-8PS VIP helicopters. *Yuriy Kirsanov*

Sultan Air

Sultan Air [KJ/SUT] was a charter airline founded in July 1989 in partnership with the Swedish airline Transwede. Although it operated almost exclusively Western types (a Sud-Est Caravelle and two Boeing 737-200s), a single An-24RV was leased from Russian carrier Bykovo-Avia in April-June 1993. The airline's operating licence was revoked in November 1993.

An-24RV TC-JUZ (ex-Bykovo Avia RA-46636, c/n 37308902) looked resplendent in the Sultan Air livery. Later the aircraft passed to Mas Air and was reregistered TC-MOC. *Yefim Gordon archive*

Turkmenistan

Turkmenistan Airlines

The **Turkmenistan Airlines** Concern was set up in 1990 as the successor of Aeroflot's Turkmen CAD. It consisted of three divisions – Akhal, Lebap and Khazar. **Akhal Aircompany [T5/AKH]** based in Ashgabat took over from the Ashkhabad UAD, operating ten An-24Bs and four An-24RVs (165th Flight), eight IL-76TDs, three Tu-154B-1s and nine Tu-154B-2s (the trijets belonged to the 369th Flight) and four Yak-42Ds, as well as three Mi-6 heavy-lift helicopters and 19 Mi-8s (166th Flight). **Khazar Aircompany [–/KHR]** based in Turkmenbashi (formerly Krasnovodsk) was the successor of the Krasnovodsk UAD with 19 An-2s, five An-24Bs, three An-24RVs and five An-26s (the 360th Flight operated the turboprops). Finally, **Lebap Aircompany [–/LEB]** based in Chardzhou – the former Chardzhou UAD – had 30 An-2s, ten Yak-40s (the latter were operated by the 167th Flight), eleven Mi-2s and 14 Mi-8s (443rd Flight). An-24Bs and An-24RVs were also stationed at Tashauz and Maryy respectively.

In 1999 the three constituent divisions of the concern merged into **Turkmenistan Airlines [T5/TUA]**. In keeping with the new policy pursued by the often-unaccountable then-President Saparmurad Niyazov, nearly all of the airline's Russian aircraft have been phased out and replaced by Western ones. Akhal was the first operator of the Boeing 737 in the CIS, receiving its first 737-300 in 1992.

Turkmenistan Airlines Tu-154B-2 EZ-85507 (c/n 81A507) is about to touch down on runway 14L at Moscow-Domodedovo. Originally this aircraft had an experimental version of the Aeroflot livery with a blue/white tail, the Turkmenistan logo being applied on the white band instead of the Soviet flag. *Yuriy Kirsanov*

IL-76TD EZ-F424 (c/n 1033418592, f/n 9008) was one of eight operated by Turkmenistan Airlines. Four of these remained in service by 2006 but this aircraft was not one of them, being stored at Ashgabat. *Yuriy Kirsanov*

EZ-26633 (c/n 5910) was one of the carrier's three An-26s; it received the Turkmenistan Airlines titles and logo in 1993 while still being registered CCCP-26633. Note that the aircraft lacks the port side navigator's observation blister, despite not being an An-26B. *Martin Novak*

United Arab Emirates

Lotus Airways

Lotus Airways based at Sharjah operated this civil-configured An-12BP (ex-Atlas Air EL-ALA, c/n 00347405) since 1998. Originally the aircraft retained the Liberian registration but was reregistered 3C-ZZD in 1999. *Sergey and Dmitriy Komissarov collection*

United Kingdom

HeavyLift

In 1992 **HeavyLift Cargo Airlines [NP/HLA]** based at London-Stansted and London-Southend started a partnership with the Russian cargo carrier Volga-Dnepr Airlines. Hence two IL-76TDs leased from this carrier received full HeavyLift colours, while several An-124-100 Ruslan 'big lifters' jointly operated by the partners wore a hybrid Volga-Dnepr/HeavyLift livery until the partnership ended in February 2001. A further IL-76TD was leased from another Russian airline, Airstan, in 1997-2000, along with two An-12BPs leased from Balkan Bulgarian Airlines in 1995-96 and the Pulkovo Air Enterprise in 1996.

IL-76TD RA-76401 (c/n 1023412399, f/n 8510) leased from Volga-Dnepr in full HeavyLift colours in 1995 completes the landing run at Moscow-Vnukovo, having just passed the intersection of runways 02-20 (foreground) and 06-24. IL-76TD 'Falsie' RA-76758 was also leased in the same livery. *Yuriy Kirsanov*

The Ukraine

Air Ukraine

An-24RV UR-47297 (c/n 07306610) demonstrates one of Air Ukraine's many pre-1993 liveries with a blue tail logo and 'Avialinii Ookrainy' titles on the rear fuselage. The aircraft was operated by the Kiev United Air Detachment based at Kiev-Zhulyany. *Martin Novak*

The Ukrainian CAD, which was one of the largest and most important in the Aeroflot system, became **Air Ukraine/Avialiniï Ukraïny [6U/UKR]**, the original Ukrainian flag carrier, after the break-up of the USSR. It included numerous flight detachments: the Borispol' UAD at Kiev-Borispol' operating Tu-134s and Tu-154s, the Kiev UAD at Kiev-Zhulyany operating An-24s and An-30s, the Chernovtsy UAD operating Let L-410s, the Dnepropetrovsk UAD operating An-26s and Yak-40s, the Donetsk UAD operating Yak-42s, the Khmel'nitskiy UAD operating L-410s, the Kirovograd UAD operating An-2s and Yak-40s, the Khar'kov UAD operating An-26s and Tu-134s, the L'vov UAD operating An-24s and Yak-42s, the Lugansk UAD operating An-2s, the Odessa UAD operating Tu-154s and Yak-40s, the Rovno UAD operating An-2s, the Simferopol' UAD operating An-24s, the Zaporozhye UAD operating Yak-40s and so on. The large fleet included dozens of An-2s, at least 20 L-410UVPs, 30 An-24Bs, 30 An-24RVs, 22 An-26s, nine An-30s, seven IL-62Ms, 15 Tu-134As, eight Tu-134AKs, 11 Tu-154Bs, four Tu-154B-1s, 17 Tu-154B-2s, four Tu-154Ms, 25 Yak-40s and 28 Yak-42s. Additionally, transport aircraft (An-12s and IL-76MDs) were occasionally leased from the Ukrainian Air Force and operated in Air Ukraine colours. Air Ukraine performed scheduled and charter services, as well as utility work.

In the course of the ten years that followed, most of Air Ukraine's territorial divisions became independent air carriers, such as Air Kharkov, Kirovograd-Avia and Rivne Universal Avia. The original Air Ukraine thus shrank to the Kiev division, which was now facing stiff competition both from the former constituent enterprises and from new ones, such as Aerosvit. It was declared bankrupt by its owner, the Ukrainian government, in December 2002 and was being restructured. There were plans to merge Air Ukraine, Aerosvit and Ukraine International Airlines as a new Air Ukraine and flag carrier. The plans came to nothing, and on 27th April 2004 Air Ukraine suspended operations, ceding the flag carrier role to Ukraine International Airlines with its all-Boeing 737 fleet.

Another An-24RV, UR-47824 (c/n 17307205), painted according to the fleetwide standard of 1993. The aircraft was built as an An-24B and upgraded to An-24RV standard during an overhaul. It was sold to ARP 410 Airlines in 2002 and later resold to Ukraine National Airlines. *Yuriy Kirsanov*

The airline's Tu-134s also showed a variety of colours. Tu-134A-3 UR-6575 (c/n [93]62215, f/n 5108) based at Kiev-Borispol' had a blue tail logo, white undersurfaces and Ukrainian-only titles; note that the 'Tu-134A-3' nose titles are applied in yellow rather than the usual white. *Yuriy Kirsanov*

Tu-134A UR-65765 (c/n [93]62315, f/n 5206) was also based at Kiev-Borispol' but wore a different tail logo with a yellow background and Ukrainian/English titles. *Dmitriy Petrochenko*

Seen on short finals to Moscow-Vnukovo, Tu-134A UR 65077 (c/n [73]60028, f/n 4203) wears the 1993-standard scheme with bilingual titles. The aircraft was refitted as a Tu-134A 'Salon' with no rear entry door and leased as such to Kiev-based TransAGO. *Yuriy Kirsanov*

Kiev-based Tu-134A-3 UR-65135 (c/n [83]60648, f/n 4808), another regular visitor to Moscow-Vnukovo, wore this striking special livery for a while in early 2001. *Yuriy Kirsanov*

Tu-154B-2 UR-85368 (c/n 79A368) operated by the Kiev division flew in this basic Aeroflot livery with a yellow tail logo and bilingual titles. It ended its days as a hulk at Kiev-Borispol' in the late 1990s.
Yuriy Kirsanov

UR-85707 (c/n 91A882) was one of only four Tu-154Ms operated by Air Ukraine. As shown here, it gained the 1993-standard livery, except that the titles were in English only. By 2006 it had been placed in storage.
Dmitriy Petrochenko

Air Ukraine operated 18 Yak-40 feederliners, including UR-87806 (c/n 9231223) operated by the Zaporozhgye United Air Detachment. The aircraft wore the yellow-badged variety of the original livery with English-only titles. In 2002 the Zaporozhye division became a separate enterprise called Constanta Airlines with which UR-87806 served on. *Sergey Popsuyevich*

Landing gear caught in mid-retraction as it departs from Moscow-Vnukovo, Yak-42D UR-42366 (c/n 4520421814047, f/n 0110) received the 1993-standard livery before the airline's Donestk division became a new entity called Donbass Airlines. *Yuriy Kirsanov*

ARP 410 Airlines

ARP 410 Airlines [–/URP] were founded in May 1999 as the flying division of the Kiev Aircraft Repair Plant No 410 located at Kiev-Zhulyany. Originally intended to perform supply functions for the parent company, it now flies passenger and cargo charters as well. The current fleet comprises one An-2, two An-24Bs, eight An-24RVs, three An-26Bs, as well as four Ka-26 helicopters, an Agusta A109C, a Bell 407 and a Bell 430. The two An-30s converted to An-30-100 executive aircraft are no longer in the fleet, and a second An-2 has been donated to the Ukrainian National Aviation Museum.

UR-26591 (c/n 14001) was one of several An-26Bs operated by ARP 410 Airlines, which bought the machine from the Russian carrier Chukotavia in 2000. Note the small Antonov logo on the nose. *Dmitriy Petrochenko*

An-24B UR-PWA (c/n 67302608) displays full ARP 410 colours as it passes overhead at Kiev-Zhulyany. The aircraft was bought in 2003 from the Ukrainian airline ICAR, which had operated it since 1998; before that it had seen service with AVL Arkhangel'sk Airlines as RA-46820. Most of ARP 410 Airlines' aircraft have Soviet-style numeric registrations. *Sergey Popsuyevich*

In contrast, An-24RV UR-CDY (c/n 47309305) wears a most unusual livery. This is probably a leftover from a lease to an unidentified Sudanese carrier that had operated the machine as ST-SHE; before that the aircraft had been registered UR-46661.
Sergey Popsuyevich

An-30 UR-30044 (c/n 0906) taken over from the Kiev division of Air Ukraine was converted to An-30-100 executive configuration. Here it is shown after arriving at Zhukovskiy for the MAKS-2001 airshow. *Dmitriy Petrochenko*

Aero-Charter

Aero-Charter Airlines [DW/UCR] based at Kiev-Borispol' have seven Yak-40s, all but two of them in executive configuration, and a single An-26. Yak-40 UR-LUX (c/n 9541542) in full colours was previously operated as UR-87230. *Sergey Popsuyevich*

Some of the airline's Yak-40s, such as UR-DWC (ex-Sirius Air UR-87987, c/n 9541144), have smart but anonymous corporate-style colour schemes. *Sergey Popsuyevich*

Aeromost

Established in 2002, Khar'kov-based **Aeromost [HT/AHW]** (= Aero Bridge) is one of the youngest Ukrainian air carriers. Its fleet consists of three An-140 twin-turboprop regional airliners, two of which wear a striking 'trumpeter angel' livery, and a single An-24B. A fourth An-140 was tragically lost in a fatal crash during a promotional trip to Iran where the An-140 is produced under licence.

Aeromost's An-24B UR-CAO (c/n 97305007), which previously served with Air Kharkov, wears this non-standard livery. Note the round badge of the InterAmi company (which outfitted the aircraft's interior) on the nose and the badge of the Air Bridge Group (the lessor) further aft. *Sergey Popsuyevich*

Aerosvit

Established on 25th March 1994 and based at Kiev-Borispol, AeroSvit Ukrainian Airlines [VV/AEW] ('AeroWorld' in Ukrainian) ' is one of the best air carriers in today's Ukraine. It operates scheduled services to 14 destinations in Russia, Poland, Serbia, Israel, Turkey, Bulgaria, Greece, Cyprus, Hungary and others, as well as domestic services and international charters. Although the airline has operated Boeing 737s since its first days and has added a couple of 767s to the fleet in 2002 and 2004, the airline occasionally uses Soviet types for charter flights. Here, An-24RV UR-46833 (c/n 17306710) leased from Crimea Airlines is seen with Aerosvit titles (in the original rendering, 'Aerosweet') and logo applied as stickers over the owner's basic green/white colours. *Sergey Popsuyevich*

Aerostar

Aerostar [–/UAR] is another Yak-40 operator based at Kiev's two airports (Zhulyany and Borispol'). Six examples are currently on strength, including Yak-40K UR-87961 (c/n 9820458) refitted to VIP configuration. *Yuriy Kirsanov*

AeroVis Airlines

Established in 2003, AeroVis Airlines [–/VIZ] based at Rovno is a cargo carrier with a fleet consisting of six An-12s, some of which are more than 30 years old. An-12BP UR-CBG (c/n 6343705) is seen here making a typically smoky take-off on a sunny winter morning; prior to that, ground personnel had had to clean away a thick layer of snow. Until the autumn of 2003 UR-CBG had been operated by Shovkovyy Shlyakh (= Silk Way) in the same basic colours. *Eero Saarela*

Antonov Airlines

Antonov Airlines [–/ADB] was established in 1989 when the Antonov design bureau (ANTK Antonov) entered into an agreement with the British cargo carrier Air Foyle to market Antonov An-124 cargo charters worldwide. This led Antonov (100% owners of the airline) to obtain its own operating certificate. True to its name, the carrier operates exclusively Antonov aircraft from the company's flight test facility (Kiev-Gostomel) which has now received airport status. The fleet included three An-12s (one has now been written off in an accident), the world's sole commercially operated An-22A, one An-26, one An-74T executive aircraft, seven An-124-100 Anteys (including the prototype of the upgraded An-124-100M-150) and the huge An-225 Mriya. The airline has earned worldwide recognition as a specialist in transporting extra heavy and outsize loads and is regularly called upon by international organisations, including NATO, to address their transport needs.

Irkutsk-built An-12AP UR-21510 (c/n 0901404) taxies in after a flight; the Nos 1 and 4 engines are already shut down to save fuel. It is now one of two in the fleet; a third An-12 was damaged beyond repair in a hard landing at the airline's home base, Kiev-Gostomel'. *Eero Saarela*

Antonov Airlines operate a single An-26 with the equally non-standard registration UR-13395 (c/n 2605). It is shown here in the static park of the fourth Aviasvit-XXI airshow which took place at Kiev-Gostomel' in June 2006.
Yefim Gordon

An-124-100 UR-82029 (c/n 19530502630, f/n 02-10) wore this interim 'hybrid' livery combining full tail colours and 'International Cargo Transporter' titles with an Aeroflot cheatline for a while.
Yefim Gordon archive

Another An-124-100, UR-82007 (c/n 19530501005, f/n 01-05) in full Antonov Airlines colours, poses for a dramatic air-to-air shot as it cruises above the clouds.
ANTK Antonov

Antonov Airlines are the sole commercial operator of the An-22. Looking magnificent in the carrier's livery, An-22A UR-09307 (c/n 043481244, f/n 0509) is shown here in the static park of the fourth Aviasvit-XXI airshow on 8th June 2006. Note the non-standard smaller chin radome housing an ROZ-1 radar instead of the usual Initsiativa radar. *Yefim Gordon*

The fleet includes a single An-74T, UR-74010 (c/n 365.470.30.450, f/n 0404), converted to an executive jet with additional cabin windows, an airstair door and additional starboard front and port rear emergency exits. *Right, Yefim Gordon; below, Sergey Popsuyevich*

Aviant

The Aviant State Aircraft Factory [–UAK] in Kiev runs a small aircraft fleet of its own – all of them Antonov types built at the plant. These include two An-24s. An-24RV-100 UR-26175 (c/n 77310810A) with upgraded avionics and a modern cabin is seen here landing on runway 30 at Zhukovskiy as it arrives for the MAKS-2005 airshow. *Yefim Gordon*

Aviant operated two An-26s and 15 An-32s, including four An-32P Firekiller waterbombers. One of them, UR-48086 (c/n 2901), is seen here giving a demonstration of its fire-fighting capabilities at an airshow. The aircraft is still in old colours with a yellow tail logo and 'Firekiller' titles on the water tanks. *Yefim Gordon*

Here the same aircraft is depicted at Kiev-Svyatoshino, Aviant's factory airfield, in the operator's current colours during the third Aviasvit-XXI in September 2004; the badge near the entry door is the Kiev city crest. As of now, two of the four An-32Ps have been sold to Libya. *Sergey Komissarov*

Atlant-SV

Simferopol'-based Atlant-SV [–/ATG], a commercial division of the Ukrainian Air Force, operated various types, including the An-12BP, IL-62M and Tu-154M. However, it was the IL-76MD that formed the backbone of the fleet. UR-76408 (c/n 0053460820, f/n 4605), one of 44 operated over the years, is seen here being unloaded at a foreign airport. The airline ceased operations in 1997, the fleet being redistributed between other operators. *Yefim Gordon archive*

Bukovyna Airlines

In 2000 the Chernovtsy division of Air Ukraine became a separate airline called Bukovyna Airlines [–/CHV] after the region of the Ukraine where the city of Chernovtsy is located. The airline operated two Tu-134A-3s, UR-65089 and UR-65790; the latter (c/n [03]63100, f/n 5601) is seen here taking off at Vnukovo. *Yuriy Kirsanov*

Crimea Air

Crimea Airlines (Aviakompaniya Krym) [MY/CRF, later OR/CRF] based in Simferopol', the 'capital' of the Crimea Peninsula, started operations on 4th October 1996, flying international and domestic scheduled and charter flights. Ten An-24RVs and two An-26s formed the backbone of the fleet, augmented by four leased Tu-154s. Two Yak-42Ds bought from Russia were operated in 1998-2000. Since 2003 the carrier operates as Krym [–KYM] with a fleet of three An-24RVs.

One of Crimea Airlines' two Yak-42s, UR-42330 (c/n 4520422505122, f/n 0806), departs Moscow-Vnukovo on a flight to Simferopol'. The aircraft was formerly RA-42330 with Simbirsk Aero; in 2002 it returned to the Russian register, now being operated by Yak Service as a VIP jet. *Yuriy Kirsanov*

As it taxies out for take-off at Moscow/Vnukovo-1 on 29th May 2000, An-24RV UR-46675 (c/n 47309607) shows the Ukrainian-language 'Krym' titles on the port side; the titles on the other side are in English. *Dmitriy Komissarov*

Dniproavia

Dniproavia [Z6/UDN] started life as the Dnepropetrovsk division of Air Ukraine. The current fleet consists of six Yak-40 feederliners (some of the original six have been retired and replaced) and five Yak-42Ds remaining of the original eight; the four An-26 freighters have been disposed of. In recent years the airline has augmented its fleet with US-built aircraft – a Boeing 737-400 and a Bell 407 helicopter. Here Yak-40 UR-88203 (c/n 9630549) taxies out for take-off at Kiev-Zhulyany. *Yefim Gordon archive*

Yak-42D UR-42426 (c/n 4520423304016, f/n 0416) soaks up the last rays of the setting sun on the east apron at Moscow-Domodedovo, showing the Ukrainian titles to starboard; those on the port side are in English. The livery is a simple adaptation of the 1993-standard Air Ukraine colours. *Dmitriy Petrochenko*

DonbassAero

In 1998 the Donetsk division of Air Ukraine became a separate entity called **Donbass – Eastern Ukrainian Airlines [7D/UDC]**. ('Donbass' is shorthand for the Donetsk Basin – a major coal mining area around Donetsk. The original fleet consisted of five An-24Bs, two An-24RVs, and 12 Yak-42 trijets. In 2003 the carrier was reorganised and rebranded as DonbassAero, retaining the original flight codes but adopting a totally new livery. In 2005-06 two of the trijets were retired, giving way to two second-hand Airbus A320s.

DonbassAero has scheduled domestic services to Chernovtsy, Dnepropetrovsk, Khar'kov, Kiev, Odessa and Simferopol' and international services to Aleppo, Athens, Baku, Beirut, Dubai, Istanbul, Larnaca, Moscow, Sharm-El-Sheikh, Tbilisi, Tel Aviv, Riga, Vilnius and Yerevan. The airline also assures several weekly cargo connections from Faro Airport, Portugal to Mauritania.

Yak-42D UR-42372 (c/n 4520423914266, f/n 1010) exemplifies the early livery worn by Donbass – Eastern Ukraine Airlines aircraft. Note the old-style red logo. *Yuriy Kirsanov*

An-24B UR-46416 (c/n 87304101) in basic Air Ukraine colours wore the new yellow/blue logo introduced in 2001. Note that the titles on the port side are in Ukrainian (Donbas – Skhidni Avialinii Ookrainy). This aircraft was retired by 2004. *Yuriy Kirsanov*

Gorlitsa Airlines

Gorlitsa Airlines [–/GOR] (the name means ‘turtle dove’) operated a single An-26B, UR-GLS (ex-Komiavia RA-26107, c/n 10109) from Kiev-Zhulyany in 1999-2001. The aircraft was on lease from Avialeasing. *Dmitriy Petrochenko*

ICAR

ICAR [C3/IPR] (the name is Russian for Icarus but also an acronym for Independent Carrier) based in Khar’kov operated two An-24s (including An-24RV UR-98104, c/n 67302310), an An-12 and a Let L-410UVP. Plans to buy five An-140 turboprops were foiled by ICAR’s bankruptcy in 2002. *Sergey Popsuyevich*

Il’yich-Avia

Il’yich-Avia [–/ILL] is a subsidiary of the Mariupol’ Steel Foundry named after Vladimir Il’yich Lenin. The fleet comprises two An-140-100s and two Yak-40s; UR-87440 *Il’yichovets 1* (ex-Bylina RA-87440, c/n 9431635) shown here still has the basic livery it wore when operated by AeroTex. *Sergey Popsuyevich*

Khors

Named after the sun god of the pagan Slavic people of ancient Rus, Khors [–/KHO] operated a succession of IL-76MDs in the 1990s (of which none remain), plus four An-12s. An-12BP UR-11332 (ex-T9-CAD, c/n 4342202) is one of the three currently on strength. *Sergey Popsuyevich*

Another An-12BP, UR-TSI (ex-Antonov Airtrack UR-UAA, c/n 6344701), in a white livery with a larger tail logo. The three An-12s are the carrier's only Soviet-built aircraft, the remainder of the fleet being McDonnell Douglas DC-9-51s and MD-82/MD-83s. *Dmitriy Petrochenko*

Lviv Airlines

Lviv Airlines [5V/UKW] were formed from the L'viv division of Air Ukraine in 1999, taking over seven An-24Bs, seven Yak-42s, two An-12s and two IL-76MDs. A single IL-18D (actually a demilitarised IL-20M electronic intelligence aircraft) was added to the fleet that year. The carrier operates domestic and international passenger and cargo services from Lviv International Airport, which is also a Ukrainian Air Force base.

One of Lviv Airlines' Yak-42s, UR-42317 (c/n 4520422202039, f/n 0505) taxies out for take-off, its ownership revealed only by the tail logo representing a very stylised L. *Eero Saarela*

Caught on approach to Moscow-Vnukovo, An-24B UR-46326 (c/n 97305503) wears basic Air Ukraine colours and Ukrainian-language titles. Some sister ships, such as UR46305, have a simple grey/white paint job. *Yefim Gordon archive*

Although the nose titles read 'IL-18d', UR-BXD is no ordinary airliner but a former IL-20M electronic intelligence aircraft stripped of its mission equipment (c/n 172011401). The two windows instead of the usual three in the forward cabin are the giveaway. *Yefim Gordon archive*

Med-Air

Med-Air took delivery of An-24RV UR-46838 (c/n 17306805), a former Air Ukraine machine, in November 1999. Note the rectangular cabin windows obtained by riveting appropriately shaped inserts into the circular window apertures during an overhaul. An An-26 freighter was also in the fleet. *Yefim Gordon archive*

Motor-Sich

Motor-Sich Airlines [M9/MSI] are a division of the Zaporozhye-based Motor-Sich aero engine factory. The current fleet consists of two An-24RVs, two Yak-40s, an An-140, a Yak-42D, two An-12s, an An-74 and two Mi-8 helicopters. Four Mi-17s and three An-26s have been sold off. Here, An-24RV UR-BXC (ex-UR-46636, c/n 37308902) shows the carrier's current livery.
Sergey Krivchikov

Seen here on finals to Moscow-Vnukovo, in 2001, Yak-40K UR-88219 (c/n 9630550) wears the old colours with Ukrainian-language titles.
Yuriy Kirsanov

In contrast, Yak-40 UR-87215 (c/n 9510540) wears the current Motor-Sich livery.
Yuriy Kirsanov

Odessa Airlines

The Odessa division of Air Ukraine based at Odessa-Tsentral'nyy also became an independent enterprise called **Odessa Airlines/Odes'ki Avialiniï [5K/ODS]** on 29th July 1996. Its fleet included eight Tu-154Bs, four Tu-154B-1 and four Yak-40s. By the turn of the century all of the Tu-154s had been retired as time-expired; the Yak-40s serve on, though the oldest one has been retired and replaced by a newer example. The Yaks have been augmented by a pair of brand-new An-140s; though technically not the launch customer for the type, Odessa Airlines was the first carrier to actually take delivery and put the An-140 into service. The airline has scheduled services to destinations in Russia, Syria, Turkey and Germany and also performs crop-spraying work with three An-2s. Odessa Airlines was privatised in 2003.

Tu-154B UR-85116 (c/n 75A116) seen here approaching Moscow-Vnukovo wore faded Air Ukraine colours with Odessa Airlines titles and logo at the closing stage of its flying career. *Yuriy Kirsanov*

Basking in the sunshine at Kiev-Zhulyany, Yak-40 UR-87624 (c/n 9130519) is a very early aircraft, as indicated by the bullet fairing at the top of the fin (it belongs to the first Yak-40 batch to have the centre engine thrust reverser). This aircraft is no longer in service. *Yefim Gordon archive*

Seen at the same location in 2005, Yak-40 UR-85327 (c/n 9330630) wears Odessa Airlines' current livery. Before that, it was operated in the black and gold basic colours of ex-owner Ukrainian National Airlines. *Sergey Popsuyevich*

Tavria-MAC

Over the years, Tavria-MAC [BE/TVM] based at Simferopol'-Zavodskoye operated a constantly shifting fleet which included four An-24RVs, an An-26, an ex-Latvian Tu-154B-2 and an ex-Romanian IL-18D (the latter was leased and subsequently bought from the Moldovan carrier Renan). Here An-24RV UR-47362 (c/n 073411330, f/n 0902) is seen on short finals to Moscow-Vnukovo in 1997; it became RA-47362 with Ryazan'aviatrans in 1999. Tavria-MAC ceased operations in 2003. *via Viktor Kravchenko*

Tavrey Aircompany

Also based at Odessa-Tsentral'nyy airport, Tavrey Aircompany [T6/TVR] operates three Yak-42Ds, all of which are re-export aircraft from Lithuania. UR-42337 (formerly Strezhevoy Airlines RA-42337, c/n 4520423606235, f/n 0507) still wears the faded basic colours of Lithuanian Airlines with whom it served as LY-AAM. *Sergey Sergeyev*

Prestige Avia

Business charter operator TransAGO [–/AKO] was founded in 1995, operating from Kiev-Borispol' – initially with a Tu-134A 'Salon' (UR 65077) and an L-410UVP leased from Air Ukraine. In 1997 the airline purchased two Tu-134A 'Salons' of its own from Lithuania; the following year it was renamed Prestige-Avia [–/PGE]. Here UR-65023 (ex-LY-ABB, c/n [63]48415, f/n 3508) is seen on approach to Moscow-Vnukovo. The carrier ceased operations in 2000. *Dmitriy Petrochenko*

South Airlines

South Airlines/Pivdenni Avialinii [YG/OTL] based at Odessa operate a fleet of six L-410UVPs, one Yak-40, three An-24s, one An-30A-100, one Yak-42D and one Tu-134A-3. The latter, UR-BZY (ex-ER-TCH, c/n 6348565, f/n 3701), shows its striking livery as it takes off from Kiev-Borispol' in 2005. *Sergey Popsuyevich*

Ukraine Air Alliance

Ukraine Air Alliance [–/UKL], a cargo carrier based at Kiev, operates two An-26Bs, a single An-32B and five IL-76TDs. One of the latter, demilitarised IL-76TD 'Falsie' UR-CAP (ex-Atlant-SV UR-76398, ex-CCCP-76674, c/n 0063466989, f/n 5008) is seen here undergoing an engine change at Sharjah, which is not only a staging area but also a maintenance centre for Soviet-built aircraft operated in the region. The aircraft was sold to East/West Cargo as ST-EWD in 2005 but transferred to Air West when the former airline had to shut down. *Aidan Curley*

Ukrainian Cargo Airways

Ukrainian Cargo Airways (UCA) [6Z/UKS] was established in July 1997 and started operations in January 1998, using decommissioned aircraft of the Ukrainian Air Force. Aircraft operated included four An-12s, two An-26s, at least 15 IL-76, two IL-78s converted to freighters, one Tu-154B-2 and three Mi-8MTV-1 helicopters.

Since 1998 UCA have been used to fulfil UN programmes in former Yugoslavia (Kosovo) and Africa (Sierra Leone, Sudan and Democratic Republic of the Congo) delivering relief cargo, transporting peacekeeping troops and evacuating refugees.

Seen here taxying at Helsinki-Vantaa, An-12BK UR-UCK (ex-Veteran Airlines 11304, c/n 9346905) differs from most of UCA's aircraft in having a blue cheatline and a grey belly. *Eero Saarela*

Demilitarised IL-76TD 'Falsie' UR-UCA (ex-IL-76MD UR-76715, c/n 0073479394, f/n 6009) was in the static park of the fourth Aviasvit-XXI airshow at Kiev-Gostomel' in June 2006, exemplifying the airline's standard grey/white colours. This aircraft is an Afghan War veteran. *Yefim Gordon*

Another Afghan War veteran with traces of built-in chaff/flare dispenser installation, IL-76MD UR-UCR (ex-UR-76728, c/n 0073475270, f/n 5708), is seen here in basic Aeroflot colours as it performs free-fall delivery of humanitarian aid. Note the small Air Ukraine titles and logo on the nose. *Sergey Popsuyevich*

Ukraine Cargo Airways' sole passenger aircraft, Tu-154B-2 UR-UCZ (ex-UR-85561, c/n 82A561), is used for carrying personnel on UN Peace Force deployments and wears non-standard 'UC Airways' titles. Note the simple boarding ramp of the kind found on military bases and industrial airfields. *Sergey Popsuyevich*

UM Air

UM Airlines [UF/UKM] (also known as Ukrainian Mediterrean Airlines or UM Air) was established in 1998, operating domestic scheduled services and services within the CIS and to the Middle East from Kiev-Borispol'. The carrier started out with two leased Tu-134As; the fleet was gradually expanded to include a single AN-24Rv, four McDonnell Douglas DC-9-51s, an MD-83 and an Airbus A320. A single Yak-42D was also operated; tragically, it crashed in May 2003.

Resplendent in full UM Air colours, Tu-134A UR-65089 *Yaroslav* (c/n [73]60180, f/n 4308) is seen here at Malta-Luqa. Sister ship UR-65076 was in basic Air Ukraine colours with the UM Air logo on a white circle. *Sergey and Dmitriy Komissarov collection*

UES Avia

UES-Avia [–/UES], a subsidiary of the United Energy Systems of the Ukraine corporation, operated a single IL-76TD 'Falsie' bought from HUK Hungarian-Ukrainian Airlines in 1996; however, the freighter was promptly leased back to the same carrier. Later UES-Avia purchased three Yak-40Ks from CSA Czech Airlines, refitting them as executive jets. One of them, UR-87591 (ex-OK-HEQ, c/n 9741056), is shown here visiting Moscow/Vnukovo-1 on 29th May 2000 with the original 'octopus' tail logo but with small Russian titles on the nose instead of the original 'United Energy Systems' titles above the windows. By September 2001 it had gained a new swallow tail logo identical to the one painted on the nose. *Dmitriy Komissarov*

Volare

Volare Aviation Enterprise [F7/VRE] was officially registered on 1st August 1994, operating initially from Krivoy Rog; later the airline moved its main base to Nikolayev. Over the years the carrier operated three An-12BPs, four An-12BKs (three remain), eight IL-76MDs (five remain now, some of which have been converted to IL-76TDs) and a single L-410UVP executive transport. Volare operates air cargo flights within Ukraine and other CIS republics, as well as Europe, the Middle East, the Far East and Africa, using a broad web of customers, agents and brokers worldwide.

Volare Aircompany An-12BK UR-BWM (ex-Ukrainian Air Force '20 Blue', c/n 00347004) touches down at Helsinki-Vantaa – nosewheels first. *Eero Saarela*

Yuzhmashavia

Yuzhmashavia [–/UMK], the flying division of the Southern Machinery Works (Yuzhnyy mashinostroitel'nyy zavod or Yuzhmash) in Dnepropetrovsk, has two IL-76TD 'Falsies' and three Yak-40s. One of the latter, UR-87508 (c/n 9521040), is shown here at Prague, with the titles mispainted as 'Yuzmashavia'. *Martin Novak*

Ukraina Air Enterprise

The Ukraina Air Enterprise operates two Tu-134AKs, including UR-65718 (c/n [13]63668, f/n 6108), one Tu-134A 'Salon', a Yak-40 'Salon 1st Class', an An-74TK-300D and four Mi-8MTV-1 helicopters. *Martin Novak*

Two IL-62M 'Salons', including UR-86528 (c/n 4038111), form the mainstay of the Ukraina Air Enterprise – that is, the Ukrainian government flight. Note the satellite communications blister above the wing leading edge. *Upper: Yuriy Kirsanov; lower: Yefim Gordon archive*

USA

Mercy Air

Despite specialising in air ambulance work with a fleet of helicopters, Mercy Air of Rialto, California also operated this smartly painted An-2 registered N75AN (ex-LY-ACE, c/n 1G211-41). The aircraft is known to have operated with the Honduran registration HR-ARR while wearing the same titles. *Yefim Gordon archive*

Uzbekistan

Uzbekistan Airways

Uzbekistan Airways (Uzbekiston Havo Yullari) [HY/UZB], the Uzbek flag carrier, was formed in 1992 by Presidential decree. Uzbekistan Airways took over the assets of the Uzbek CAD's Tashkent UAD (160th, 203rd and 219th Flights) and Samarkand UAD (163rd Flight). The passenger aircraft inherited from Aeroflot (and Soviet-designed aircraft delivered afterwards) included 19 An-24Bs, four An-24RVs, eight IL-62Ms, ten IL-86s, two IL-114s, one Tu-154B, five Tu-154B-1s, 13 Tu-154B-2s, three Tu-154Ms and 25 Yak-40s. Cargo operations were performed with two An-12BPs and 21 IL-76TDs.

A strong presence on the international air transport market was planned from the outset, as indicated by the airline's inaugural service being from Tashkent to London, rather than a domestic route. When international routes became top priority, Airbus A310s were leased, starting in 1993. The international fleet now uses a mix of A310s, Boeing 757-200s, Boeing 767-300s and Avro RJ85s, the Tupolevs and the Ilyushin jet airliners having been withdrawn. However, Uzbekistan Airways operate the Pratt & Whitney-powered IL-114-100 (for which the carrier was the launch customer) and plan to buy more as a replacement for the ageing An-24s and Yak-40s, supporting the national aircraft industry.

The IL-86 was the largest aircraft in Uzbekistan Airways' fleet. UK 86090 (c/n 51483207061), the youngest of the ten operated, is seen here on final approach to Moscow-Sheremet'yevo. The type has now been withdrawn. *Yuriy Kirsanov*

Pictured taxying at Moscow-Domodedovo, Tu-154B-3 UK 85344 (c/n 79A344) wears the early standard livery with Uzbekistan titles. *Dmitriy Petrochenko*

IL-62M UK 86574 (ex-Interflug D-AOAH, ex-DDR-SEO, c/n 3344833) taxies to the gate at Frankfurt-Main. Like the IL-86s, the long-haul IL-62s are up for sale. *Robert J Ruffle archive*

An-24B UK 46387 (c/n 07306110) taxies out for take-off at Tashkent-Yoozhnyy past Luftwaffe C160 Transalls used by the Western coalition forces combating the Taliban militia in neighbouring Afghanistan. The type is due for replacement with the locally manufactured IL-114-100. *Mir Aviatsii*

One of the original two IL-114s powered by Russian Klimov TV7-117S engines, UK 91001 (c/n 1013823024, f/n 0106), is seen at Zhukovskiy on 3rd September 1993 during the MAKS-93 airshow. This aircraft is now in service with Vyborg Airlines as RA-91014; the Uzbeks have expressed a preference for the Pratt & Whitney-powered IL-114-100. *Dmitriy Komissarov*

Still lacking the UK prefix, IL-62M 86579 (ex-Interflug D-AOAN, ex-DDR-SEZ, c/n 2951636) has a non-standard livery with a white tail. Note the IL-62's distinctive feature – the telescopic twin-wheel tail support extended when the aircraft is parked. *Robert J Ruffle archive*

Kicking up a local snowstorm, IL-62M UK 86575 (ex-Interflug D-AOAJ, ex-DDR-SES, c/n 1647928) displays the carrier's current livery. In July 2005 the aircraft was sold to Moscow-based Interavia, becoming RA-86575. *Dmitriy Petrochenko*

Uzbekistan Airways operated two An-12BPs, UK 11369 and UK 11372, in the 1990s. One of them stayed in service long enough to receive the current livery in national flag colours. *Dmitriy Petrochenko*

IL-76TD UK 76448 (c/n 1023413443, f/n 8701) is shown here in the old white-tailed livery. The titles on the other side read Uzbekiston Havo Yullari. *Yuriy Kirsanov*

Sister ship UK 76449 (c/n 1003403058, f/n 7705) taxies at Moscow-Domodedovo, showing the livery worn by most of the carrier's IL-76TDs. Both aircraft were operated by the Russian airline East Line in the late 1990s. *Dmitriy Petrochenko*

A fine shot of IL-76TD UK 76358 (c/n 1023410339, f/n 8405), tucking up its landing gear as it climbs away from Moscow-Domodedovo. *Yuriy Kirsanov*

UK 76793 (c/n 0093498951, f/n 7408) is the only known IL-76TD to wear the 'blue top/green belly' livery. *Yefim Gordon archive*

Avialeasing

Avialeasing Aviation Co [AD/TWN, later EC/TWN] was established in 1992 as an Uzbek-US joint venture – the first private airline in Uzbekistan, operating at least seven IL-76TDs. None of these remain in the fleet now; currently the carrier operates three An-12BPs and five An-26s from Tashkent and Miami-Opa Locka. Two of the An-26s have even been placed on the US register.

An-26B UK 26002 (ex-TAPO-Avia UK 93914, c/n 9104) bought in 2000 sits parked at Tashkent in full Avialeasing colours. In 2004 it was sold to the USA, becoming N8038Y with SRX Transcontinental. Oddly, all the while another An-26 existed in Russia as CCCP-26002 and later RA-26002. *Dmitriy Petrochenko*

The cheatline on IL-76TD UK 76410 (c/n 1023412411, f/n 8603) reveals previous ownership by the Kazakh carrier Sayakhat which operated the freighter as UN 76410 until 1997. The machine was later repainted in the standard colours and eventually sold to AZAL Cargo as 4K-AZ16 in 2001. *Dmitriy Petrochenko*

TAPO Avia

TAPO-Avia [PQ/CTP, later 4C/TPR] is the flying division of the Tashkent Aircraft Production Corporation (TAPO). Its fleet consists mainly of IL-76TDs, one of which, UK 76375 (c/n 1033414496, f/n 8804) is seen here seconds before touching down on runway 14L at Moscow-Domodedovo. *Yuriy Kirsanov*

TAPO-Avia also has two An-12s, including An-12BP UK 58644 (c/n 2340303) which is pictured here awaiting the next flight at Moscow-Domodedovo on 3rd November 1998. The airline also had an IL-114T freighter but, tragically, it crashed fatally at Domodedovo on 4th December 1999. *Dmitriy Komissarov*

Uzbekistan Government

The Uzbekistan government flight operated this Tu-154B-2 'Salon', UK 85600 (ex-85600, ex-Afghan Government YA-TAT, c/n 84A600). The colour scheme is similar to Uzbekistan Airways' current livery, except for the white tail, white titles and the Uzbek coat-of-arms on the nose. *Yuriy Kirsanov*

IL-62M 'Salon' UK 86569 (c/n 1356234) wore the same colours. This aircraft was reportedly sold to the Gambian government flight, becoming C5-RTG (that is, Republic of the Gambia). Currently the presidential jet role in Uzbekistan is filled by a Boeing 757-23P. *Yuriy Kirsanov*

Vietnam Airlines

Founded in 1951 as Air Vietnam, the airline was renamed Hang Khong Viet Nam [VN/HVN] and became an operator of Soviet aircraft after the Vietnam War. The fleet included such types as the An-24B, IL-14, IL-18D, Tu-134A/AK/B and Yak-40. An example of the latter type, VN-A446 (c/n 9631748), is seen above: it survived the ranaming of the carrier to Vietnam Airlines in 1990 but was retired in 1998. *via Chris Knott*

Vietnam Airlines had four Tu-134B-3s. VN-A120 (c/n [43]66360, f/n 6362) was the only one delivered in non-VIP configuration; it crashed in Phnom Penh on 3rd September 1997.
Yuriy Kirsanov

Vietnam Air Services Co

Vietnam Air Services Co (VASCO) [0V/VFC] is a subsidiary of Vietnam Airlines tasked with crop-spraying and photo survey work. Lately, however, it has leased two freighters (including an An-12) and two An-38-120 feederliners, including RA-41902 (c/n 4163847010002) leased from Vostok Airlines.
Yefim Gordon archive

Yemen

Yemenia

The Yemen Air Force took delivery of three IL-76TDs which were jointly operated with the nation's flag carrier Yemenia [IY/IYE]. 7O-ADG delivered in November 1995 (ex-RA-76405, c/n 1023412402, f/n 8601) is seen here visiting Sharjah in 2001. *Sergey Popsuyevich*

Zaire (Congo-Kinshasa)

Gloria Airways

Soviet commercial aircraft had been delivered to Zaire before the 1997 revolution which led to a change of name to Congo-Kinshasa – Blue Airlines had a single An-26 and two An-28s, while the Air Transport Office leased an An-12 and an IL-18. In recent years An-12s, An-26s and An-28s have seen service with several Congolese carriers. Here, Gloria Airways An-26 9Q-CMV (ex-RA-26090, c/n 11506) is 'unbuttoned' for maintenance at Sharjah. *Sergey Popsuyevich*

We hope you enjoyed this book . . .

Midland Publishing titles are edited and designed by an experienced and enthusiastic team of specialists.

We always welcome ideas from authors or readers for books they would like to see published.

In addition, our associate, Midland Counties Publications, offers an exceptionally wide range of aviation, military, naval and transport books and videos for sale by mail-order worldwide.

For a copy of the appropriate catalogue, or to order further copies of this book, and any other Midland Publishing titles, please write, telephone, fax or e-mail to:

Midland Counties Publications
4 Watling Drive, Hinckley,
Leics, LE10 3EY, England
Tel: (+44) 01455 254 450
Fax: (+44) 01455 233 737
E-mail: midlandbooks@compuserve.com
www.midlandcountiessuperstore.com

US distribution by Specialty Press – see page 2.

RUSSIAN AIRLINES AND THEIR AIRCRAFT

Dmitriy Komissarov & Yefim Gordon

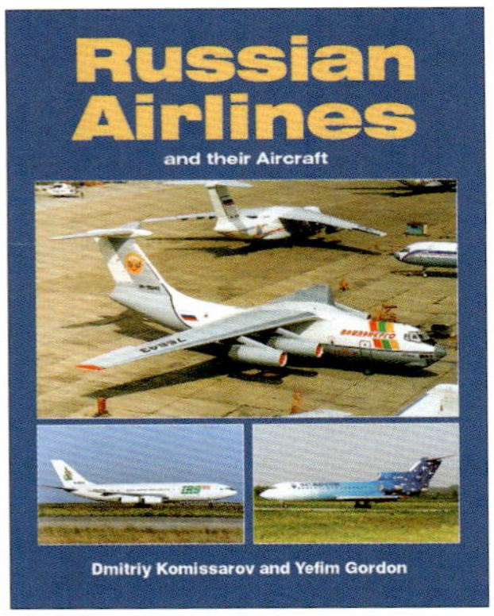

Following the ending of Aeroflot's monopoly in 1992 and the break-up of the Soviet Union, the Russian civil air transport scene has been considerably transformed.

This full-colour album covers the major airlines operating in Russia today, illustrating the types operated by each carrier, their equipment and the various colour schemes worn by them. A brief history and fleet information are provided for each airline, as are detailed photo captions.

Softback, 280 x 215 mm, 160 pages
449 full colour photographs
978 1 85780 176 7 **£19.99**

AIRLINERS OF THE 1970s

Gerry Manning

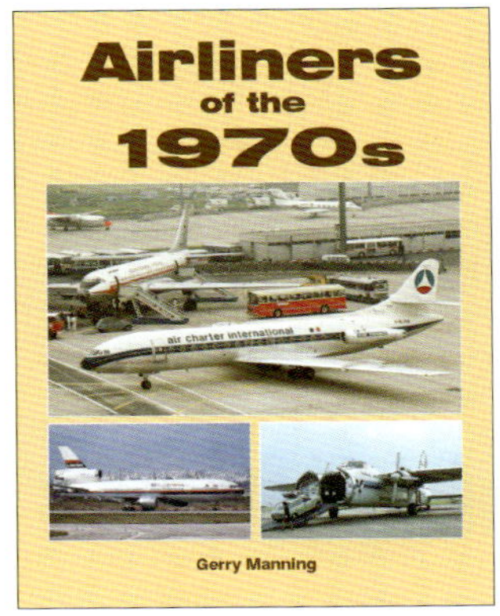

In a highly pictorial look at a decade which saw much change in the world air travel scene, well-travelled aviation photographer Gerry Manning has assembled an exciting collection of images from all over the globe. Over 60 different types are featured, from the propliners still hard at work to the newly-introduced Concorde and Tupolev Tu-144 supersonic transports by way of the first Boeing 747 services and introduction of the first Airbus product: the A300.

Softback, 280 x 215 mm, 144 pages
311 colour photographs
978 1 85780 213 9 **£18.99**

EUROPEAN AIRLINES

John K Morton

This book portrays with extended captions, a broad selection of over 100 primarily passenger-carrying airlines from 33 European countries. Starting in Ireland and continuing around the continent, the reader is taken on a 'tour' of European nations, and in each case liveries of based airlines, including all the major flag carriers, scheduled and charter airlines are included. As wide a variety of types – jets and propliners – as possible are shown and an equally-varied range of colour schemes.

Softback, 280 x 215 mm, 112 pages
188 full colour photographs
978 1 85780 210 8 **£14.99**

Aerofax ILYUSHIN IL-18/20/22

A Versatile Turboprop Transport

Yefim Gordon and Dmitriy Komissarov

The IL-18 four-turboprop airliner first flew in 1957 and was supplied to many 'friendly nations' in Eastern Europe, Asia, Africa, Middle East and the Caribbean. Its uses included passenger and cargo, VIP transportation, support of Antarctic research stations, electronic espionage and various research programmes. All versions are described, as are many test and development aircraft, the IL-20M ELINT, IL-20RT space tracker, IL-22 airborne command post, IL-24N for ice reconnaissance and IL-38 ASW aircraft.

Softback, 280 x 215 mm, 160 pages
184 b/w, 67 colour photos, plus dwgs
978 1 85780 157 6 **£19.99**

Aerofax TUPOLEV Tu-154

The USSR's Medium-Range Jet Airliner

Dmitriy Komissarov

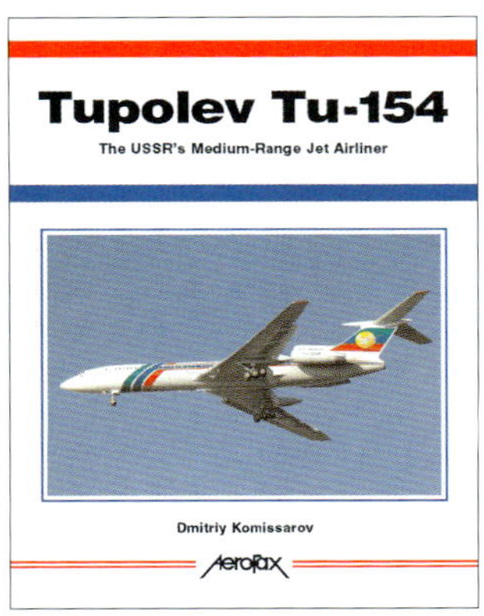

Sharing the rear-engine, 'T-tail' layout of the Boeing 727 and the Hawker Siddeley Trident, the prototype Tu-154 took to the air on 3rd October 1968.

The type was to become not only the principal medium-haul aircraft of the Soviet airline Aeroflot but also one of the Soviet Union's bestsellers on the commercial aircraft market.

A detailed production list, operator lists and accident details are included, as are numerous colour photographs and line drawings.

Sbk, 280 x 215 mm, 192pp,
269 col, 71 b/w photos, plus dwgs
978 1 85780 241 2 **£19.99**

Red Star Volume 24 TUPOLEV Tu-144

Russia's Concorde

Yefim Gordon and Vladimir Rigmant

Tasked with creating a supersonic transport ahead of the West as a matter of national prestige, Andrey Tupolev met the target at the cost of a tremendous research and development effort. The Tu-144 took to the air in December 1968, ahead of the Anglo-French Concorde. This detailed account includes the reasons behind its premature withdrawal and a description of its recent use in a joint research programme with NASA.

Softback, 280 x 215 mm, 128 pages
151 b/w photos, 15 pages of colour plus drawings
978 1 85780 216 0 **£19.99**

Red Star Volume 31 TUPOLEV Tu-114

The First Soviet Intercontinental Airliner

Yefim Gordon & Vladimir Rigmant

Based on the Tu-95 bomber, the Tu-114 was the largest airliner of its time, with capacity for up to 220 passengers. It was able to fly at speeds similar to those of jet aircraft, and set a number of records, including the speed record for a turbo-prop aircraft that still stands 50 years later. 31 Tu-114s served with Aeroflot, the last being withdrawn in 1975. However, a number were subsequently converted into AWACS aircraft as the Tu-126 'Moss'.

Sbk, 280 x 215 mm, 128 pages
194 b/w, 51 colour photos,
plus line drawings
978 1 85780 246 7 **£19.99**

Russian Airliners Outside Russia

Some 400 colour photographs, with extended and informative captions, form a comprehensive guide to the changing usage of Russian airliners in the post-Soviet era. Whilst many of the passenger airliners formerly operated by client states have been supplanted by Western-built aircraft, Russian aircraft remain an important part of the ad hoc freight charter scene and dominate niche markets like the carriage of outsize cargo.

MIDLAND
An imprint of
Ian Allan Publishing

www.ianallanpublishing.com

ISBN 978-1-85780-252-8

9 781857 802528

USA $36.95 UK £19.99

Printed in England